A MATHEMATICIAN'S APOLOGY

A MATHEMATICIAN'S APOLOGY

BY

G. H. HARDY

WITH A FOREWORD BY

C. P. SNOW

CAMBRIDGE
UNIVERSITY PRESS

University Printing House, Cambridge CB2 8BS, United Kingdom

One Liberty Plaza, 20th Floor, New York, NY 10006, USA

477 Williamstown Road, Port Melbourne, VIC 3207, Australia

314-321, 3rd Floor, Plot 3, Splendor Forum, Jasola District Centre, New Delhi - 110025, India

79 Anson Road, #06-04/06, Singapore 079906

Cambridge University Press is part of the University of Cambridge.

It furthers the University's mission by disseminating knowledge in the pursuit of education, learning and research at the highest international levels of excellence.

www.cambridge.org
Information on this title: www.cambridge.org/9781107604636

First edition 1940
Reprinted 1941, 1948
Reprinted with Foreword
by C. P. Snow 1967, 1969,
1973, 1976, 1977, 1979, 1981,
1982, 1984, 1985, 1987, 1988,
1989, 1990
Canto edition 1992
26th printing 2019

A catalogue record for this publication is available from the British Library

ISBN 978-1-107-60463-6 Paperback

To

JOHN LOMAS

who asked me to write it

FOREWORD

IT was a perfectly ordinary night at Christ's high table, except that Hardy was dining as a guest. He had just returned to Cambridge as Sadleirian professor, and I had heard something of him from young Cambridge mathematicians. They were delighted to have him back: he was a *real* mathematician, they said, not like those Diracs and Bohrs the physicists were always talking about: he was the purest of the pure. He was also unorthodox, eccentric, radical, ready to talk about anything. This was 1931, and the phrase was not yet in English use, but in later days they would have said that in some indefinable way he had star quality.

So, from lower down the table, I kept studying him. He was then in his early fifties: his hair was already grey, above skin so deeply sunburnt that it stayed a kind of Red Indian bronze. His face was beautiful—high cheek bones, thin nose, spiritual and austere but capable of dissolving into convulsions of internal gamin-like amusement. He had opaque brown eyes, bright as a bird's—a kind of eye not uncommon among those with a gift for conceptual thought. Cambridge at that time was full of unusual and distinguished faces—

but even then, I thought that night, Hardy's stood out.

I do not remember what he was wearing. It may easily have been a sports coat and grey flannels under his gown. Like Einstein, he dressed to please himself: though, unlike Einstein, he diversified his casual clothing by a taste for expensive silk shirts.

As we sat round the combination-room table, drinking wine after dinner, someone said that Hardy wanted to talk to me about cricket. I had been elected only a year before, but Christ's was then a small college, and the pastimes of even the junior fellows were soon identified. I was taken to sit by him. I was not introduced. He was, as I later discovered, shy and self-conscious in all formal actions, and had a dread of introductions. He just put his head down as it were in a butt of acknowledgment, and without any preamble whatever began:

'You're supposed to know something about cricket, aren't you?' Yes, I said, I knew a bit.

Immediately he began to put me through a moderately stiff viva. Did I play? What sort of performer was I? I half-guessed that he had a horror of persons, then prevalent in academic society, who devotedly studied the literature but had never played the game. I trotted out my credentials, such as they were. He appeared to find the reply partially reassuring, and went on to more tactical questions. Whom should I have chosen as captain for the last test match a year before (in 1930)? If the selectors had decided that Snow was the man to save England, what would have been my strategy and tactics? ('You are allowed to act, if you are sufficiently modest, as non-playing captain.') And so on, oblivious to the rest of the table. He was quite absorbed.

As I had plenty of opportunities to realize in the future, Hardy had no faith in intuitions or impressions, his own or anyone else's. The only way to assess someone's knowledge, in Hardy's view, was to examine him. That went for mathematics, literature, philosophy, politics, anything you like. If the man had bluffed and then wilted under the questions, that was his lookout. First things came first, in that brilliant and concentrated mind.

That night in the combination-room, it was necessary to discover whether I should be tolerable as a cricket companion. Nothing else mattered. In the end he smiled with immense charm, with child-like openness, and said that Fenner's (the university cricket ground) next season might be bearable after all, with the prospect of some reasonable conversation.

Thus, just as I owed my acquaintanceship with Lloyd George to his passion for phrenology, I owed my friendship with Hardy to having wasted a disproportionate amount of my youth on cricket. I don't know what the moral is. But it was a major piece of luck for me. This was intellectually the most valuable friendship of my life. His mind, as I have just mentioned, was brilliant and concentrated: so much so that by his side anyone else's seemed a little muddy, a little pedestrian and confused. He wasn't a great genius, as Einstein and Rutherford were. He said, with his usual clarity, that if the word meant anything he was not a genius at all. At his best, he said, he was for a short time the fifth best pure mathematician in the world. Since his character was as beautiful and candid as his mind, he always made the point that his friend and collaborator Littlewood was an appreciably more powerful mathematician than he was, and that his protégé Ramanujan really had natural genius in the sense (though not to the extent, and nothing like so effectively) that the greatest mathematicians had it.

People sometimes thought he was under-rating himself, when he spoke of these friends. It is true that he was magnanimous, as far from envy as a man can be: but I think one mistakes his quality if one doesn't accept his judgment. I prefer to believe in his own statement in *A Mathematician's Apology*, at the same time so proud and so humble:

'I still say to myself when I am depressed and

find myself forced to listen to pompous and tiresome people, "Well, I have done one thing you could never have done, and that is to have collaborated with Littlewood and Ramanujan on something like equal terms."'

In any case, his precise ranking must be left to the historians of mathematics (though it will be an almost impossible job, since so much of his best work was done in collaboration). There is something else, though, at which he was clearly superior to Einstein or Rutherford or any other great genius: and that is at turning any work of the intellect, major or minor or sheer play, into a work of art. It was that gift above all, I think, which made him, almost without realizing it, purvey such intellectual delight. When *A Mathematician's Apology* was first published, Graham Greene in a review wrote that along with Henry James's notebooks, this was the best account of what it was like to be a *creative artist*. Thinking about the effect Hardy had on all those round him, I believe that is the clue.

He was born, in 1877, into a modest professional family. His father was Bursar and Art Master at Cranleigh, then a minor public (English for private) school. His mother had been senior mistress at the Lincoln Training College for teachers. Both were gifted and mathematically inclined. In his case, as in that of most mathematicians, the gene pool doesn't need searching for. Much of his childhood, unlike Einstein's, was typical of a future mathematician's. He was demonstrating a formidably high I.Q. as soon as, or before, he learned to talk. At the age of two he was writing down numbers up to millions (a common sign of mathematical ability). When he was taken to church he amused himself by factorizing the numbers of the hymns: he played with numbers from that time on, a habit which led to the touching scene at Ramanujan's sick-bed: the scene is well known, but later on I shall not be able to resist repeating it.

It was an enlightened, cultivated, highly literate Victorian childhood. His parents were probably a little obsessive, but also very kind. Childhood in such a Victorian family was as gentle a time as anything we could provide, though probably intellectually somewhat more exacting. His was unusual in just two respects. In the first place, he suffered from an acute self-consciousness at an unusually early age, long before he was twelve. His parents knew he was prodigiously clever, and so did he. He came top of his class in all subjects. But, as the result of coming top of his class, he had to go in front of the school to receive prizes: and that he could not bear. Dining with me one night, he said that he deliberately used to try to get his answers wrong so as to be spared this intolerable ordeal. His capacity for dissimulation, though, was always minimal: he got the prizes all the same.

Some of this self-consciousness wore off. He became competitive. As he says in the *Apology*: 'I do not remember having felt, as a boy, any *passion* for mathematics, and such notions as I may have had of the career of a mathematician were far from noble. I thought of mathematics in terms of examinations and scholarships: I wanted to beat other boys, and this seemed to be the way in which I could do so most decisively.' Nevertheless, he had to live with an over-delicate nature. He seems to have been born with three skins too few. Unlike Einstein, who had to subjugate his powerful ego in the study of the external world before he could attain his moral stature, Hardy had to strengthen an ego which wasn't much protected. This at times in later life made him self-assertive (as Einstein never was) when he had to take a moral stand. On the other hand, it gave him his introspective insight and beautiful candour, so that he could speak of himself with absolute simplicity (as Einstein never could).

I believe this contradiction, or tension, in his temperament was linked with a curious tic in his behaviour. He was the classical anti-narcissist. He could not endure having his photograph taken: so far as I know, there are only five snapshots in existence. He would not have any looking glass in his rooms, not even a shaving mirror. When he went to a hotel, his first action was to cover all the looking-glasses with towels. This would have been odd enough, if his face had been like a gargoyle: superficially it might seem odder, since all his life he was good-looking quite out of the ordinary. But, of course, narcissism and anti-narcissism have nothing to do with looks as outside observers see them.

This behaviour seems eccentric, and indeed it was. Between him and Einstein, though, there was a difference in kind. Those who spent much time with Einstein—such as Infeld—found him grow stranger, less like themselves, the longer they knew him. I am certain that I should have felt the same. With Hardy the opposite was true. His behaviour was often different, bizarrely so, from ours: but it came to seem a kind of superstructure set upon a nature which wasn't all that different from our own, except that it was more delicate, less padded, finer-nerved.

The other unusual feature of his childhood was more mundane: but it meant the removal of all practical obstacles throughout his entire career. Hardy, with his limpid honesty, would have been the last man to be finicky on this matter. He knew what privilege meant, and he knew that he had possessed it. His family had no money, only a schoolmaster's income, but they were in touch with the best educational advice of late nineteenth-century England. That particular kind of information has always been more significant in this country than any amount of wealth. The scholarships have been there all right, if one knew how to win them. There was never the slightest chance of the young Hardy being lost—as there was of the young Wells or the young Einstein. From the age of twelve he had only to survive, and his talents would be looked after.

At twelve, in fact, he was given a scholarship at Winchester, then and for long afterwards the best mathematical school in England, simply on the strength of some mathematical work he had done at Cranleigh. (Incidentally, one wonders if any great school could be so elastic nowadays?) There he was taught mathematics in a class of one: in classics he was as good as the other top collegers. Later, he admitted that he had been well-educated, but he admitted it reluctantly. He disliked the school, except for its classes. Like all Victorian public schools, Winchester was a pretty rough place. He nearly died one winter. He envied Littlewood in his cared-for home as a day boy at St Paul's or other friends at our free-and-easy grammar schools. He never went near Winchester after he had left it: but he left it, with the inevitability of one who had got on to the right tramlines, with an open scholarship to Trinity.

He had one curious grievance against Winchester. He was a natural ball-games player with a splendid eye. In his fifties he could usually beat the university second string at real tennis, and in his sixties I saw him bring off startling shots in the cricket nets. Yet he had not had an hour's coaching at Winchester: his method was defective: if he had been coached, he thought, he would have been a really good batsman, not quite first-class, but not too far away. Like all his judgments on himself, I believe that one is quite true. It is strange that, at the zenith of Victorian games-worship, such a talent was utterly missed. I suppose no one thought it worth looking for in the school's top scholar, so frail and sickly, so defensively shy.

It would have been natural for a Wykehamist of his period to go to New College. That wouldn't have made much difference to his professional career (though, since he always liked Oxford better than Cambridge, he might have stayed there all his life, and some of us would have missed a treat). He decided to go to Trinity instead, for a reason that he describes, humorously but with his usual undecorated truth, in the *Apology*. 'I was about fifteen when (in a rather odd way) my ambitions took a sharper turn. There is a book by "Alan St Aubyn" (actually Mrs

Frances Marshall) called *A Fellow of Trinity*, one of a series dealing with what is supposed to be Cambridge college life... There are two heroes, a primary hero called Flowers, who is almost wholly good, and a secondary hero, a much weaker vessel, called Brown. Flowers and Brown find many dangers in university life... Flowers survives all these troubles, is Second Wrangler and succeeds automatically to a Fellowship (as I suppose he would have done then). Brown succumbs, ruins his parents, takes to drink, is saved from delirium tremens during a thunderstorm only by the prayers of the Junior Dean, has much difficulty in obtaining even an Ordinary Degree, and ultimately becomes a missionary. The friendship is not shattered by these unhappy events, and Flowers's thoughts stray to Brown, with affectionate pity, as he drinks port and eats walnuts for the first time in Senior Combination Room.

'Now Flowers was a decent enough fellow (so far as "Alan St Aubyn" could draw one), but even my unsophisticated mind refused to accept him as clever. If he could do these things, why not I? In particular, the final scene in Combination Room fascinated me completely, and from that time, until I obtained one, mathematics meant to me primarily a Fellowship of Trinity.'

Which he duly obtained, after getting the highest place in the Mathematical Tripos Part II, at the age of 22. On the way, there were two minor vicissitudes. The first was theological, in the high Victorian manner. Hardy had decided—I think before he left Winchester—that he did not believe in God. With him, this was a black-and-white decision, as sharp and clear as all other concepts in his mind. Chapel at Trinity was compulsory. Hardy told the Dean, no doubt with his own kind of shy certainty, that he could not conscientiously attend. The Dean, who must have been a jack-in-office, insisted that Hardy should write to his parents and tell them so. They were orthodox religious people, and the Dean knew, and Hardy knew much more, that the news would give them pain—pain such as we, seventy years later, cannot easily imagine.

Hardy struggled with his conscience. He wasn't worldly enough to slip the issue. He wasn't even worldly enough—he told me one afternoon at Fenner's, for the wound still rankled—to take the advice of more sophisticated friends, such as George Trevelyan and Desmond MacCarthy, who would have known how to handle the matter. In the end he wrote the letter. Partly because of that incident, his religious disbelief remained open and active ever after. He refused to go into any college chapel even for formal business, like electing a master. He had clerical friends, but God was his personal enemy. In all this there was an echo of the nineteenth century: but one would be wrong, as always with Hardy, not to take him at his word.

Still, he turned it into high jinks. I remember, one day in the thirties, seeing him enjoy a minor triumph. It happened in a Gentlemen *v.* Players match at Lord's. It was early in the morning's play, and the sun was shining over the pavilion. One of the batsmen, facing the Nursery end, complained that he was unsighted by a reflection from somewhere unknown. The umpires, puzzled, padded round by the sight-screen. Motor-cars? No. Windows? None on that side of the ground. At last, with justifiable triumph, an umpire traced the reflection down—it came from a large pectoral cross reposing on the middle of an enormous clergyman. Politely the umpire asked him to take it off. Close by, Hardy was doubled up in Mephistophelian delight. That lunch time, he had no leisure for eating: he was writing postcards (postcards and telegrams were his favourite means of communication) to each of his clerical friends.

But in his war against God and God's surrogates, victory was not all on one side. On a quiet and lovely May evening at Fenner's, round about the same period, the chimes of six o'clock fell across the ground. 'It's rather unfortunate,' said Hardy simply, 'that some of the happiest hours of my life should have been spent within sound of a Roman Catholic church.'

The second minor upset of his undergraduate years was professional. Almost since the time of Newton, and all through the nineteenth century, Cambridge had been dominated by the examination for the old Mathematical Tripos. The English have always had more faith in competitive examinations than any other people (except perhaps the Imperial Chinese): they have conducted these examinations with traditional justice: but they have often shown remarkable woodenness in deciding what the examinations should be like. That is, incidentally, true to this day. It was certainly true of the Mathematical Tripos in its glory. It was an examination in which the questions were usually of considerable mechanical difficulty—but unfortunately did not give any opportunity for the candidate to show mathematical imagination or insight or any quality that a creative mathematician needs. The top candidates (the Wranglers —a term which still survives, meaning a First Class) were arranged, on the basis of marks, in strict numerical order. Colleges had celebrations when one of their number became Senior Wrangler: the first two or three Wranglers were immediately elected Fellows.

It was all very English. It had only one disadvantage, as Hardy pointed out with his polemic clarity, as soon as he had become an eminent mathematician and was engaged, together with his tough ally Littlewood, in getting the system abolished: it had effectively ruined serious mathematics in England for a hundred years.

In his first term at Trinity, Hardy found himself caught in this system. He was to be trained as a racehorse, over a course of mathematical exercises which at nineteen he knew to be meaningless. He was sent to a famous coach, to whom most potential Senior Wranglers went. This coach knew all the obstacles, all the tricks of the examiners, and was sublimely uninterested in the subject itself. At this point the young Einstein would have rebelled: he would either have left Cambridge or done no formal work for the next three years. But Hardy was born into the more intensely professional English climate (which has its merits as well as its demerits). After considering changing his subject to history, he had the sense to find a real mathematician to teach him. Hardy paid him a tribute in the *Apology*: 'My eyes were first opened by Professor Love, who taught me for a few terms and gave me my first serious conception of analysis. But the great debt which I owe to him—he was, after all, primarily an applied mathematician—was his advice to read Jordan's famous *Cours d'Analyse*: and I shall never forget the astonishment with which I read that remarkable work, the first inspiration for so many mathematicians of my generation, and learned for the first time as I read it what mathematics really meant. From that time onwards I was in my way a real mathematician, with sound mathematical ambitions and a genuine passion for mathematics.'

He was Fourth Wrangler in 1898. This faintly irritated him, he used to confess. He was enough of a natural competitor to feel that, though the race was ridiculous, he ought to have won it. In 1900 he took Part II of the Tripos, a more respect-worthy examination, and got his right place and his Fellowship.

From that time on, his life was in essence settled. He knew his purpose, which was to bring rigour into English mathematical analysis. He did not deviate from the researches which he called 'the one great permanent happiness of my life'. There were no anxieties about what he should do. Neither he nor anyone else doubted his great talent. He was elected to the Royal Society at thirty-three.

In many senses, then, he was unusually lucky. He did not have to think about his career. From the time he was twenty-three he had all the leisure that a man could want, and as much money as he needed. A bachelor don in Trinity in the 1900's was comfortably off. Hardy was sensible about money, spent it when he felt impelled (sometimes for singular purposes, such as fifty-mile taxi-rides), and otherwise was not at all unworldly about investments. He played his games and indulged his eccentricities. He was living in some of the best intellectual company in the world—G. E. Moore, Whitehead, Bertrand Russell, Trevelyan, the high Trinity society which was shortly to find its artistic complement in Bloomsbury. (Hardy himself had links with Bloomsbury, both of personal friendship and of sympathy.) In a brilliant circle, he was one of the most brilliant young men—and, in a quiet way, one of the most irrepressible.

I will anticipate now what I shall say later. His life remained the life of a brilliant young man until he was old: so did his spirit: his games, his interests, kept the lightness of a young don's. And, like many men who keep a young man's interests into their sixties, his last years were the darker for it.

Much of his life, though, he was happier than most of us. He had a great many friends, of surprisingly different kinds. These friends had to pass some of his private tests: they needed to possess a quality which he called 'spin' (this is a cricket term, and untranslatable: it implies a certain obliquity or irony of approach: of recent public figures, Macmillan and Kennedy would get high marks for spin, Churchill and Eisenhower not). But he was tolerant, loyal, extremely high-spirited, and in an undemonstrative way fond of his friends. I once was compelled to go and see him in the morning, which was always his set time for mathematical work. He was sitting at his desk, writing in his beautiful calligraphy. I murmured some commonplace politeness about hoping that I wasn't disturbing him. He suddenly dissolved into his mischievous grin. 'As you ought to be able to notice, the answer to that is that you are. Still, I'm usually glad to see you.' In the sixteen years we knew each other, he didn't say anything more demonstrative than that: except on his deathbed, when he told me that he looked forward to my visits.

I think my experience was shared by most of his close friends. But he had, scattered through his life, two or three other relationships, different in kind. These were intense affections, absorbing, non-physical but exalted. The one I knew about was for a young man whose nature was as spiritually delicate as his own. I believe, though I only picked this up from chance remarks, that the same was true of the others. To many people of my generation, such relationships would seem either unsatisfactory or impossible. They were neither the one nor the other; and, unless one takes them for granted, one doesn't begin to understand the temperament of men like Hardy (they are rare, but not as rare as white rhinoceroses), nor the Cambridge society of his time. He didn't get the satisfactions that most of us can't help finding: but he knew himself unusually well, and that didn't make him unhappy. His inner life was his own, and very rich. The sadness came at the end. Apart from his devoted sister, he was left with no one close to him.

With sardonic stoicism he says in the *Apology*—which for all its high spirits is a book of desperate sadness—that when a creative man has lost the power or desire to create—'It is a pity but in that case he does not matter a great deal anyway, and it would be silly to bother about him.' That is how he treated his personal life outside mathematics. Mathematics was his justification. It was easy to forget this, in the sparkle of his company: just as it was easy in the presence of Einstein's moral passion to forget that to himself his justification was his search for the physical laws. Neither of those two ever forgot it. This was the core of their lives, from young manhood to death.

Hardy, unlike Einstein, did not make a quick start. His early papers, between 1900 and 1911, were good enough to get him into the Royal Society and win him an international reputation: but he did not regard them as important. Again, this wasn't false modesty: it was the judgment of a master who knew to an inch which of his work had value and which hadn't.

In 1911 he began a collaboration with Littlewood which lasted thirty-five years. In 1913 he discovered Ramanujan and began another collaboration. All his major work was done in these two partnerships, most of it in the one with Littlewood, the most famous collaboration in the history of mathematics. There has been nothing like it in any science, or, so far as I know, in any other field of creative activity. Together they produced nearly a hundred papers, a good many of them 'in the Bradman class'. Mathematicians not intimate with Hardy in his later years, nor with cricket, keep repeating that his highest term of praise was 'in the Hobbs class'. It wasn't: very reluctantly, because Hobbs was one of his pets, he had to alter the order of merit. I once had a postcard from him, probably in 1938, saying 'Bradman is a whole class above any batsman who has ever lived: if Archimedes, Newton and Gauss remain in the Hobbs class, I have to admit the possibility of a class above them, which I find difficult to imagine. They had better be moved from now on into the Bradman class.'

The Hardy–Littlewood researches dominated English pure mathematics, and much of world pure mathematics, for a generation. It is too early to say, so mathematicians tell me, to what extent they altered the course of mathematical analysis: nor how influential their work will appear in a hundred years. Of its enduring value there is no question.

Theirs was, as I have said, the greatest of all

collaborations. But no one knows how they did it: unless Littlewood tells us, no one will ever know. I have already given Hardy's judgment that Littlewood was the more powerful mathematician of the two: Hardy once wrote that he knew of 'no one else who could command such a combination of insight, technique and power'. Littlewood was and is a more normal man than Hardy, just as interesting and probably more complex. He never had Hardy's taste for a kind of refined intellectual flamboyance, and so was less in the centre of the academic scene. This led to jokes from European mathematicians, such as that Hardy had invented him so as to take the blame in case there turned out anything wrong with one of their theorems. In fact, he is a man of at least as obstinate an individuality as Hardy himself.

At first glance, neither of them would have seemed the easiest of partners. It is hard to imagine either of them suggesting the collaboration in the first place. Yet one of them must have done. No one has any evidence how they set about it. Through their most productive period, they were not at the same university. Harald Bohr (brother of Niels Bohr, and himself a fine mathematician) is reported as saying that one of their principles was this: if one wrote a letter to the other, the recipient was under no obligation to reply to it, or even to read it.

I can't contribute anything. Hardy talked to me, over a period of many years, on almost every conceivable subject, except the collaboration. He said, of course, that it had been the major fortune of his creative career: he spoke of Littlewood in the terms I have given: but he never gave a hint of their procedures. I didn't know enough mathematics to understand their papers, but I picked up some of their language. If he had let slip anything about their methods, I don't think I should have missed it. I am fairly certain that the secrecy—quite uncharacteristic of him in matters which to most would seem more intimate—was deliberate.

About his discovery of Ramanujan, he showed no secrecy at all. It was, he wrote, the one romantic incident in his life: anyway, it is an admirable story, and one which showers credit on nearly everyone (with two exceptions) in it. One morning early in 1913, he found, among the letters on his breakfast table, a large untidy envelope decorated with Indian stamps. When he opened it, he found sheets of paper by no means fresh, on which, in a non-English holograph, were line after line of symbols. Hardy glanced at them without enthusiasm. He was by this time, at the age of thirty-six, a world famous mathematician: and world famous mathematicians, he had already discovered, are unusually exposed to cranks. He was accustomed to receiving manuscripts from strangers, proving the prophetic wisdom of the Great Pyramid, the revelations of the Elders of Zion, or the cryptograms that Bacon had inserted in the plays of the so-called Shakespeare.

So Hardy felt, more than anything, bored. He glanced at the letter, written in halting English, signed by an unknown Indian, asking him to give an opinion of these mathematical discoveries. The script appeared to consist of theorems, most of them wild or fantastic looking, one or two already well-known, laid out as though they were original. There were no proofs of any kind. Hardy was not only bored, but irritated. It seemed like a curious kind of fraud. He put the manuscript aside, and went on with his day's routine. Since that routine did not vary throughout his life, it is possible to reconstruct it. First he read *The Times* over his breakfast. This happened in January, and if there were any Australian cricket scores, he would start with them, studied with clarity and intense attention.

Maynard Keynes, who began his career as a mathematician and who was a friend of Hardy's, once scolded him: if he had read the stock exchange quotations half an hour each day with the same concentration he brought to the cricket scores, he could not have helped becoming a rich man.

Then, from about nine to one, unless he was giving a lecture, he worked at his own mathematics. Four hours creative work a day is about the limit for a mathematician, he used to say. Lunch, a light meal, in hall. After lunch he loped off for a game of real tennis in the university court. (If it had been summer, he would have walked down to Fenner's to watch cricket.) In the late afternoon, a stroll back to his rooms. That particular day, though, while the timetable wasn't altered, internally things were not going according to plan. At the back of his mind, getting in the way of his complete pleasure in his game, the Indian manuscript nagged away. Wild theorems. Theorems such as he had never seen before, nor imagined. A fraud of genius? A question was forming itself in his mind. As it was Hardy's mind, the question was forming itself with epigrammatic clarity: is a fraud of genius more probable than an unknown mathematician of genius? Clearly the answer was no. Back in his rooms in Trinity, he had another look at the script. He sent word to Littlewood (probably by messenger, certainly not by telephone, for which, like all mechanical contrivances including fountain pens, he had a deep distrust) that they must have a discussion after hall.

When the meal was over, there may have been a slight delay. Hardy liked a glass of wine, but, despite the glorious vistas of 'Alan St. Aubyn' which had fired his youthful imagination, he found he did not really enjoy lingering in the combination-room over port and walnuts. Littlewood, a good deal more *homme moyen sensuel*, did. So there may have been a delay. Anyway, by nine o'clock or so they were in one of Hardy's rooms, with the manuscript stretched out in front of them.

That is an occasion at which one would have liked to be present. Hardy, with his combination of remorseless clarity and intellectual panache (he was very English, but in argument he showed the characteristics that Latin minds have often assumed to be their own): Littlewood, imaginative, powerful, humorous. Apparently it did not take them long. Before midnight they knew, and knew for certain. The writer of these manuscripts was a man of genius. That was as much as they could judge, that night. It was only later that Hardy decided that Ramanujan was, in terms of *natural* mathematical genius, in the class of Gauss and Euler: but that he could not expect, because of the defects of his education, and because he had come on the scene too late in the line of mathematical history, to make a contribution on the same scale.

It all sounds easy, the kind of judgment great mathematicians should have been able to make. But I mentioned that there were two persons who do not come out of the story with credit. Out of chivalry Hardy concealed this in all that he said or wrote about Ramanujan. The two people concerned have now been dead, however, for many years, and it is time to tell the truth. It is simple. Hardy was not the first eminent mathematician to be sent the Ramanujan manuscripts. There had been two before him, both English, both of the highest professional standard. They had each returned the manuscripts without comment. I don't think history relates what they said, if anything, when Ramanujan became famous. Anyone who has been sent unsolicited material will have a sneaking sympathy with them.

Anyway, the following day Hardy went into action. Ramanujan must be brought to England, he decided. Money was not a major problem. Trinity has usually been good at supporting unorthodox talent (the college did the same for Kapitsa a few years later). Once Hardy was determined, no human agency could have stopped Ramanujan, but they needed a certain amount of help from a superhuman one.

Ramanujan turned out to be a poor clerk in Madras, living with his wife on twenty pounds a year. But he was also a Brahmin, unusually strict about his religious observances, with a mother who was even stricter. It seemed impossible that he could break the proscriptions and cross the water. Fortunately his mother had the highest respect for the goddess of Namakkal. One morning Ramanujan's mother made a startling announcement. She had had a dream on the previous night, in which she saw her son seated in a big hall among a group of Europeans, and the goddess of Namakkal had commanded her not to stand in the way of her son fulfilling his life's purpose. This, say Ramanujan's Indian biographers, was a very agreeable surprise to all concerned.

In 1914 Ramanujan arrived in England. So far as Hardy could detect (though in this respect I should not trust his insight far) Ramanujan, despite the difficulties of breaking the caste proscriptions, did not believe much in theological doctrine, except for a vague pantheistic benevolence, any more than Hardy did himself. But he did certainly believe in ritual. When Trinity put him up in college—within four years he became a Fellow—there was no 'Alan St. Aubyn' apolausticity for him at all. Hardy used to find him ritually changed into his pyjamas, cooking vegetables rather miserably in a frying pan in his own room.

Their association was a strangely touching one. Hardy did not forget that he was in the presence of genius: but genius that was, even in mathematics, almost untrained. Ramanujan had not been able to enter Madras University because he could not matriculate in English. According to Hardy's report, he was always amiable and good-natured, but no doubt he sometimes found Hardy's conversation outside mathematics more than a little baffling. He seems to have listened with a patient smile on his good, friendly, homely face. Even inside mathematics they had to come to terms with the difference in their education. Ramanujan was self-taught: he knew nothing of the modern rigour: in a sense he didn't know what a proof was. In an uncharacteristically sloppy moment, Hardy once wrote that if he had been better educated, he would have been less Ramanujan. Coming back to his ironic senses, Hardy later corrected himself and said that the statement was nonsense. If Ramanujan had been better educated, he would have been even more wonderful than he was. In fact, Hardy was obliged to teach him some formal mathematics as though Ramanujan had been a scholarship candidate at Winchester. Hardy said that this was the most singular experience of his life: what did modern mathematics look like to someone who had the deepest insight, but who had literally never heard of most of it?

Anyway, they produced together five papers of the highest class, in which Hardy showed supreme originality of his own (more is known of the details of this collaboration than of the Hardy–Littlewood one). Generosity and imagination were, for once, rewarded in full.

This is a story of human virtue. Once people had started behaving well, they went on behaving better. It is good to remember that England gave

Ramanujan such honours as were possible. The Royal Society elected him a Fellow at the age of thirty (which, even for a mathematician, is very young). Trinity also elected him a Fellow in the same year. He was the first Indian to be given either of these distinctions. He was amiably grateful. But he soon became ill. It was difficult, in war-time, to move him to a kinder climate.

Hardy used to visit him, as he lay dying in hospital at Putney. It was on one of those visits that there happened the incident of the taxi-cab number. Hardy had gone out to Putney by taxi, as usual his chosen method of conveyance. He went into the room where Ramanujan was lying. Hardy, always inept about introducing a conversation, said, probably without a greeting, and certainly as his first remark: 'I thought the number of my taxi-cab was 1729. It seemed to me rather a dull number.' To which Ramanujan replied: 'No, Hardy! No, Hardy! It is a very interesting number. It is the smallest number expressible as the sum of two cubes in two different ways.'

That is the exchange as Hardy recorded it. It must be substantially accurate. He was the most honest of men; and further, no one could possibly have invented it.

Ramanujan died of tuberculosis, back in Madras, two years after the war. As Hardy wrote in the *Apology*, in his roll-call of mathematicians: 'Galois died at twenty-one, Abel at twenty-seven, Ramanujan at thirty-three, Riemann at forty...I do not know an instance of a major mathematical advance initiated by a man past fifty.'

If it had not been for the Ramanujan collaboration, the 1914–18 war would have been darker for Hardy than it was. But it was dark enough. It left a wound which reopened in the second war. He was a man of radical opinions all his life. His radicalism, though, was tinged with the turn-of-the-century enlightenment. To people of my generation, it sometimes seemed to breathe a lighter, more innocent air, than the one we knew.

Like many of his Edwardian intellectual friends, he had a strong feeling for Germany. Germany had, after all, been the great educating force of the nineteenth century. To Eastern Europe, to Russia, to the United States, it was the German universities which had taught the meaning of research. Hardy hadn't much use for German philosophy or German literature: his tastes were too classical for that. But in most respects the German culture, including its social welfare, appeared to him higher than his own.

Unlike Einstein, who had a much more rugged sense of political existence, Hardy did not know much of Wilhelmine Germany at first hand. And, though he was the least vain of people, he would have been less than human if he had not enjoyed being more appreciated in Germany than in his own country. There is a pleasant anecdote, dating from this period, in which Hilbert, one of the greatest of German mathematicians, heard that Hardy lived in a not specially agreeable set of rooms in Trinity (actually in Whewell's Court). Hilbert promptly wrote in measured terms to the Master, pointing out that Hardy was the best mathematician, not only in Trinity but in England, and should therefore have the best rooms.

So Hardy, like Russell and many of the high Cambridge intelligentsia, did not believe that the war should have been fought. Further, with his ingrained distrust of English politicians, he thought the balance of wrong was on the English side. He could not find a satisfactory basis for conscientious objection; his intellectual rigour was too strong for that. In fact, he volunteered for service under the Derby scheme, and was rejected on medical grounds. But he felt increasingly isolated in Trinity, much of which was vociferously bellicose.

Russell was dismissed from his lectureship, in circumstances of overheated complexity (Hardy was to write the only detailed account of the case a quarter of a century later, in order to comfort himself in another war). Hardy's close friends were away at the war. Littlewood was doing ballistics as a Second Lieutenant in the Royal Artillery. Owing to his cheerful indifference he had the distinction of remaining a Second Lieutenant through the four years of war. Their collaboration was interfered with, though not entirely stopped. It was the work of Ramanujan which was Hardy's solace during the bitter college quarrels.

I sometimes thought he was, for once, less than fair to his colleagues. Some were pretty crazed, as men are in war-time. But some were long-suffering and tried to keep social manners going. After all, it was a triumph of academic uprightness that they should have elected his protégé Ramanujan, at a time when Hardy was only just on speaking terms with some of the electors, and not at all with others.

Still, he was harshly unhappy. As soon as he could, he left Cambridge. He was offered a chair at Oxford in 1919: and immediately walked into the happiest time of his life. He had already done great work with Ramanujan and Littlewood, but now the collaboration with Littlewood rose to its full power. Hardy was, in Newton's phrase, 'in the prime of his age for invention', and this came in his early forties, unusually late for a mathematician.

Coming so late, this creative surge gave him the feeling, more important to him than to most men, of timeless youth. He was living the young man's life which was first nature to him. He played more real tennis, and got steadily better at it (real tennis was an expensive game and took a largish slice out of a professorial income). He made a good many visits to American universities, and loved the country. He was one of the few Englishmen of his time who was fond, to an extent approximately equal, of the United States and the Soviet Union. He was certainly the only Englishman of his or any other time who wrote a serious suggestion to the Baseball Commissioners, proposing a technical emendation to one of the rules. The twenties, for him and for most liberals of his generation, was a false dawn. He thought the misery of the war was swept away into the past.

He was at home in New College as he had never been in Cambridge. The warm domestic conversational Oxford climate was good for him. It was there, in a college at that time small and intimate, that he perfected his own style of conversation. There was company eager to listen to him after hall. They could take his eccentricities. He was not only a great and good man, they realized, but an entertaining one. If he wanted to play conversational games, or real (though bizarre) games on the cricket field, they were ready to act as foils. In a casual and human fashion, they made a fuss of him. He had been admired and esteemed before, but not made a fuss of in that fashion.

No one seemed to care—it was a gossipy college joke—that he had a large photograph of Lenin in his rooms. Hardy's radicalism was somewhat unorganized, but it was real. He had been born, as I have explained, into a professional family: almost all his life was spent among the *haute bourgeoisie*: but in fact he behaved much more like an aristocrat, or more exactly like one of the romantic projections of an aristocrat. Some of this attitude, perhaps, he had picked up from his friend Bertrand Russell. But most of it was innate. Underneath his shyness, he just didn't give a damn.

He got on easily, without any patronage, with the poor, the unlucky and diffident, those who were handicapped by race (it was a symbolic stroke of fate that he discovered Ramanujan). He preferred them to the people whom he called the *large bottomed*: the description was more psychological than anatomical, though there was a famous nineteenth-century Trinity aphorism by Adam Sedgwick: 'No one ever made a success in this world without a large bottom.' To Hardy the large bottomed were the confident, booming, imperialist bourgeois English. The designation included most bishops, headmasters, judges, and all politicians, with the single exception of Lloyd George.

Just to show his allegiances, he accepted one public office. For two years (1924–26) he was President of the Association of Scientific Workers. He said sarcastically that he was an odd choice, being 'the most unpractical member of the most unpractical profession in the world'. But in important things he was not so unpractical. He was deliberately standing up to be counted. When, much later, I came to work with Frank Cousins, it gave me a certain quiet pleasure to reflect that I had had exactly two friends who had held office in the Trade Union movement, him and G. H. Hardy.

That late, not quite Indian, summer in Oxford in the twenties was so happy that people wondered that he ever returned to Cambridge. Which he did in 1931. I think there were two reasons. First and most decisive, he was a great professional. Cambridge was still the centre of English mathematics, and the senior mathematical chair there was the correct place for a professional. Secondly, and rather oddly, he was thinking about his old age. Oxford colleges, in many ways so human and warm, are ruthless with the old: if he remained at New College he would be turned out of his rooms as soon as he retired, under the age limit, from his professorship. Whereas, if he returned to Trinity, he could stay in college until he died. That is in effect what he managed to do.

When he came back to Cambridge—which was the time that I began to know him—he was in the afterglow of his great period. He was still happy. He was still creative, not so much as in the twenties, but enough to make him feel that the power was still there. He was as spirited as he had been at New College. So we had the luck to see him very nearly at his best.

In the winters, after we had become friendly, we gave each other dinner in our respective colleges once a fortnight. When summer came, it was taken for granted that we should meet at the cricket ground. Except on special occasions he still did mathematics in the morning, and did not arrive at Fenner's until after lunch. He used to walk round the cinderpath with a long, loping, clumping-footed stride (he was a slight spare man, physically active even in his late fifties, still playing real tennis) head down, hair, tie, sweaters, papers all flowing , a figure that caught everyone's eyes. 'There goes a Greek poet, I'll be bound,' once said some cheerful farmer as Hardy passed the score-board. He made for his favourite place, opposite the pavilion, where he could catch each ray of sun—he was obsessively heliotropic. In order to deceive the sun into shining, he brought with him, even on a fine May afternoon, what he called his 'anti-God battery'. This consisted of three or four sweaters, an umbrella belonging to his sister, and a large envelope containing mathematical manuscripts, such as a Ph.D. dissertation, a paper which he was refereeing for the Royal Society, or some tripos answers. He would explain to an acquaintance that God, believing that Hardy expected the weather to change and give him a chance to work, counter-suggestibly arranged that the sky should remain cloudless.

There he sat. To complete his pleasure in a long afternoon watching cricket, he liked the sun to be shining and a companion to join in the fun. Technique, tactics, formal beauty—those were the deepest attractions of the game for him. I won't try to explain them: they are incommunicable unless one knows the language: just as some of Hardy's classical aphorisms are inexplicable unless one knows either the language of cricket or of the theory of numbers, and preferably both. Fortunately for a good many of our friends, he also had a relish for the human comedy.

He would have been the first to disclaim that he had any special psychological insight. But he was the most intelligent of men, he had lived with his eyes open and read a lot, and he had obtained a good generalized sense of human nature—robust, indulgent, satirical, and utterly free from moral vanity. He was spiritually candid as few men are (I doubt if anyone could be more candid), and he had a mocking horror of pretentiousness, self-righteous indignation, and the whole stately pantechnicon of the hypocritical virtues. Now cricket, the most beautiful of games, is also the most hypocritical. It is supposed to be the ultimate expression of the team spirit. One ought to prefer to make 0 and see one's side win, than make 100 and see it lose (one very great player, like Hardy a man of innocent candour, once remarked mildly that he never managed to feel so). This particular ethos inspired Hardy's sense of the ridiculous. In reply he used to expound a counterbalancing series of maxims. Examples:

'Cricket is the only game where you are playing against eleven of the other side and ten of your own.'

'If you are nervous when you go in first, nothing restores your confidence so much as seeing the other man get out.'

If his listeners were lucky, they would hear other remarks, not relevant to cricket, as sharp-edged in conversation as in his writing. In the *Apology* there are some typical specimens: here are a few more.

'It is never worth a first class man's time to express a majority opinion. By definition, there are plenty of others to do that.'

'When I was an undergraduate one might, if one were sufficiently unorthodox, suggest that Tolstoi came within touching distance of George Meredith as a novelist: but, of course, no one else possibly could.' (This was said about the intoxications of fashion: it is worth remembering that he had lived in one of the most brilliant of all Cambridge generations.)

'For any serious purpose, intelligence is a very minor gift.'

'Young men ought to be conceited: but they oughtn't to be imbecile.' (Said after someone had tried to persuade him that *Finnegans Wake* was the final literary masterpiece.)

'Sometimes one has to say difficult things, but one ought to say them as simply as one knows how.'

Occasionally, as he watched cricket, his ball-by-ball interest flagged. Then he demanded that we should pick teams: teams of humbugs, club-men, bogus poets, bores, teams whose names began with HA (numbers one and two Hadrian and Hannibal), SN, all-time teams of Trinity, Christ's, and so on. In these exercises I was at a disadvantage: let anyone try to produce a team of world-figures whose names start with SN. The Trinity team is overwhelmingly strong (Clerk Maxwell, Byron, Thackeray, Tennyson aren't certain of places): while Christ's, beginning strongly with Milton and Darwin, has nothing much to show from number 3 down.

Or he had another favourite entertainment. 'Mark that man we met last night', he said, and someone had to be marked out of 100 in each of the categories Hardy had long since invented and defined. STARK, BLEAK ('a stark man is not necessarily bleak: but all bleak men without exception want to be considered stark'), DIM,

OLD BRANDY, SPIN, and some others. STARK, BLEAK and DIM are self-explanatory (the Duke of Wellington would get a flat 100 for STARK and BLEAK, and 0 for DIM). OLD BRANDY was derived from a mythical character who said that he never drank anything but old brandy. Hence, by extrapolation, OLD BRANDY came to mean a taste that was eccentric, esoteric, but just within the bounds of reason. As a character (and in Hardy's view, though not mine, as a writer also) Proust got high marks for OLD BRANDY: so did F. A. Lindemann (later Lord Cherwell).

The summer days passed. After one of the short Cambridge seasons, there was the University match. Arranging to meet him in London was not always simple, for, as I mentioned before, he had a morbid suspicion of mechanical gadgets (he never used a watch), in particular of the telephone. In his rooms in Trinity or his flat in St George's Square, he used to say, in a disapproving and slightly sinister tone: 'If you *fancy yourself* at the telephone, there is one in the next room.' Once in an emergency he had to ring me up: angrily his voice came at me: 'I shan't hear a word you say, so when I'm finished I shall immediately put the receiver down. It's important you should come round between nine and ten tonight.' Clonk.

Yet, punctually, he arrived at the University match. There he was at his most sparkling, year after year. Surrounded by friends, men and women, he was quite released from shyness. He was the centre of all our attention, which he didn't dislike. Sometimes one could hear the party's laughter from a quarter of the way round the ground.

In those last of his happy years, everything he did was light with grace, order, a sense of style. Cricket is a game of grace and order, which is why he found formal beauty in it. His mathematics, so I am told, had these same aesthetic qualities, right up to his last creative work. I have given the impression, I fancy, that in private he was a conversational performer. To an extent, that was true: but he was also, on what he would have called 'non-trivial' occasions (meaning occasions important to either participant) a serious and concentrated listener. Of other eminent persons whom, by various chances, I knew at the same period, Wells was, on the whole, a worse listener than one expected: Rutherford distinctly better: Lloyd George one of the best listeners of all time. Hardy didn't suck impressions and knowledge out of others' words, as Lloyd George did, but his mind was at one's disposal. When, years before I wrote it, he heard of the concept of *The Masters*, he cross-examined me, so that I did most of the talking. He produced some good ideas. I wish he had been able to read the book, which I think that he might have liked. Anyway, in that hope I dedicated it to his memory.

In the note at the end of the *Apology* he refers to other discussions. One of these was long-drawn-out, and sometimes, on both sides, angry. On the second world war we each had passionate but, as I shall have to say a little later, different opinions. I didn't shift his by one inch. Nevertheless, though we were separated by a gulf of emotion, on the plane of reason he recognized what I was saying. That was true in any argument I had with him.

Through the thirties he lived his own version of a young man's life. Then suddenly it broke. In 1939 he had a coronary thrombosis. He recovered, but real tennis, squash, the physical activities he loved, were over for good. The war darkened him still further, just as the first war had. To him they were connected pieces of lunacy, we were all at fault, he couldn't identify himself with the war—once it was clear the country would survive—any more than he had done in 1914. One of his closest friends died tragically. And—I think there is no doubt these griefs were inter-connected—his creative powers as a mathematician at last, in his sixties, left him.

That is why *A Mathematician's Apology* is, if read with the textual attention it deserves, a book of haunting sadness. Yes, it is witty and sharp with intellectual high spirits: yes, the crystalline clarity and candour are still there: yes, it is the testament of a creative artist. But it is also, in an understated stoical fashion, a passionate lament for creative powers that used to be and that will never come again. I know nothing like it in the language: partly because most people with the literary gift to express such a lament don't come to feel it: it is very rare for a writer to realize, with the finality of truth, that he is absolutely finished.

Seeing him in those years, I couldn't help thinking of the price he was paying for his young man's life. It was like seeing a great athlete, for years in the pride of his youth and skill, so much younger and more joyful than the rest of us, suddenly have to accept that the gift has gone. It is common to meet great athletes who have gone, as they call it, over the hill: fairly quickly the feet get heavier (often the eyes last longer), the strokes won't come off, Wimbledon is a place to be dreaded, the crowds go to watch someone else. That is the point at which a good many athletes take to drink. Hardy didn't take to drink: but he took to something like despair. He recovered enough physically to have ten minutes batting at the nets, or to play his pleasing elaboration (with a complicated set of bisques) of Trinity bowls. But it was often hard to rouse his interest—three or four years before his interest in everything was so sparkling as sometimes to tire us all out. 'No one should ever be bored', had been one of his axioms. 'One can be horrified, or disgusted, but one can't be bored.' Yet now he was often just that, plain bored.

It was for that reason that some of his friends, including me, encouraged him to write the story of Bertrand Russell and Trinity in the 1914–18 war. People who didn't know how depressed Hardy was thought the whole episode was now long over and ought not to be resurrected. The truth was, it enlivened him to have any kind of purpose. The book was privately circulated. It has never been obtainable by the public, which is a pity, for it is a small-scale addition to academic history.

I used such persuasion as I had to get him to write another book, which he had in happier days promised me to do. It was to be called *A Day at the Oval* and was to consist of himself watching cricket for a whole day, spreading himself in disquisitions on the game, human nature, his reminiscences, life in general. It would have been an eccentric minor classic: but it was never written.

I wasn't much help to him in those last years. I was deeply involved in war-time Whitehall, I was preoccupied and often tired, it was an effort to get to Cambridge. But I ought to have made the effort more often than I did. I have to admit, with remorse, that there was, not exactly a chill, but a gap in sympathy between us. He lent me his flat in

Pimlico—a dark and seedy flat with the St George's Square gardens outside and what he called an 'old brandy' attractiveness—for the whole of the war. But he didn't like me being so totally committed. People he approved of oughtn't to give themselves whole-heartedly to military functions. He never asked me about my work. He didn't want to talk about the war. While I, for my part, was impatient and didn't show anything like enough consideration. After all, I thought, I wasn't doing this job for fun: as I had to do it, I might as well extract the maximum interest. But that is no excuse.

At the end of the war I did not return to Cambridge. I visited him several times in 1946. His depression had not lifted, he was physically failing, short of breath after a few yards walk. The long cheerful stroll across Parker's Piece, after the close of play, was gone for ever: I had to take him home to Trinity in a taxi. He was glad that I had gone back to writing books: the creative life was the only one for a serious man. As for himself, he wished that he could live the creative life again, no better than it had been before: his own life was over.

I am not quoting his exact words. This was so unlike him that I wanted to forget, and I tried, by a kind of irony, to smear over what had just been said. So that I have never remembered precisely.

I attempted to dismiss it to myself as a rhetorical flourish.

In the early summer of 1947 I was sitting at breakfast when the telephone rang. It was Hardy's sister: he was seriously ill, would I come up to Cambridge at once, would I call at Trinity first? At the time I didn't grasp the meaning of the second request. But I obeyed it, and in the porter's lodge at Trinity that morning found a note from her: I was to go to Donald Robertson's rooms, he would be waiting for me.

Donald Robertson was the Professor of Greek, and an intimate friend of Hardy's: he was another member of the same high, liberal, graceful Edwardian Cambridge. Incidentally, he was one of the few people who called Hardy by his Christian name. He greeted me quietly. Outside the windows of his room it was a calm and sunny morning. He said:

'You ought to know that Harold has tried to kill himself.'

Yes, he was out of danger: he was for the time being all right, if that was the phrase to use. But Donald was, in a less pointed fashion, as direct as Hardy himself. It was a pity the attempt had failed. Hardy's health had got worse: he could not in any case live long: even walking from his rooms to hall had become a strain. He had made a completely deliberate choice. Life on those terms he would not endure: there was nothing in it. He had collected enough barbiturates: he had tried to do a thorough job, and had taken too many.

I was fond of Donald Robertson, but I had met him only at parties and at Trinity high table. This was the first occasion on which we had talked intimately. He said, with gentle firmness, that I ought to come up to see Hardy as often as I could: it would be hard to take, but it was an obligation: probably it would not be for long. We were both wretched. I said goodbye, and never saw him again.

In the Evelyn nursing home, Hardy was lying in bed. As a touch of farce, he had a black eye. Vomiting from the drugs, he had hit his head on the lavatory basin. He was self-mocking. He had made a mess of it. Had anyone ever made a bigger mess? I had to enter into the sarcastic game. I had never felt less like sarcasm, but I had to play. I talked about other distinguished failures at bringing off suicide. What about German generals in the last war? Beck, Stulpnagel, they had been remarkably incompetent at it. It was bizarre to hear myself saying these things. Curiously enough, it seemed to cheer him up.

After that, I went to Cambridge at least once a week. I dreaded each visit, but early on he said that he looked forward to seeing me. He talked a little, nearly every time I saw him, about death.

He wanted it: he didn't fear it: what was there to fear in nothingness? His hard intellectual stoicism had come back. He would not try to kill himself again. He wasn't good at it. He was prepared to wait. With an inconsistency which might have pained him—for he, like most of his circle, believed in the rational to an extent that I thought irrational—he showed an intense hypochondriac curiosity about his own symptoms. Constantly he was studying the oedema of his ankles: was it greater or less that day?

Mostly, though—about fifty-five minutes in each hour I was with him—I had to talk cricket. It was his only solace. I had to pretend a devotion to the game which I no longer felt, which in fact had been lukewarm in the thirties except for the pleasure of his company. Now I had to study the cricket scores as intently as when I was a schoolboy. He couldn't read for himself, but he would have known if I was bluffing. Sometimes, for a few minutes, his old vivacity would light up. But if I couldn't think of another question or piece of news, he would lie there, in the kind of dark loneliness that comes to some people before they die.

Once or twice I tried to rouse him. Wouldn't it be worth while, even if it was a risk, to go and see one more cricket match together? I was now better off than I used to be, I said. I was prepared to stand him a taxi, his old familiar means of transport, to any cricket ground he liked to name. At that he brightened. He said that I might have a dead man on my hands. I replied that I was ready to cope. I thought that he might come: he knew, I knew, that his death could only be a matter of months: I wanted to see him have one afternoon of something like gaiety. The next time I visited him he shook his head in sadness and anger. No, he couldn't even try: there was no point in trying.

It was hard enough for me to have to talk cricket. It was harder for his sister, a charming intelligent woman who had never married and who had spent much of her life looking after him. With a humorous skill not unlike his own old form, she collected every scrap of cricket news she could find, though she had never learned anything about the game.

Once or twice the sarcastic love of the human comedy came bursting out. Two or three weeks before his death, he heard from the Royal Society that he was to be given their highest honour, the Copley Medal. He gave his Mephistophelian grin, the first time I had seen it in full splendour in all those months. 'Now I know that I must be pretty near the end. When people hurry up to give you honorific things there is exactly one conclusion to be drawn.'

After I heard that, I think I visited him twice. The last time was four or five days before he died.

There was an Indian test team playing in Australia, and we talked about them.

It was in that same week that he told his sister: 'If I knew that I was going to die today, I think I should still want to hear the cricket scores.'

He managed something very similar. Each evening that week before she left him, she read a chapter from a history of Cambridge university cricket. One such chapter contained the last words he heard, for he died suddenly, in the early morning.

PREFACE

I AM indebted for many valuable criticisms to Professor C. D. Broad and Dr C. P. Snow, each of whom read my original manuscript. I have incorporated the substance of nearly all of their suggestions in my text, and have so removed a good many crudities and obscurities.

In one case I have dealt with them differently. My § 28 is based on a short article which I contributed to *Eureka* (the journal of the Cambridge Archimedean Society) early in the year, and I found it impossible to remodel what I had written so recently and with so much care. Also, if I had tried to meet such important criticisms seriously, I should have had to expand this section so much as to destroy the whole balance of my essay. I have therefore left it unaltered, but have added a short statement of the chief points made by my critics in a note at the end.

G. H. H.

18 *July* 1940

I

It is a melancholy experience for a professional mathematician to find himself writing about mathematics. The function of a mathematician is to do something, to prove new theorems, to add to mathematics, and not to talk about what he or other mathematicians have done. Statesmen despise publicists, painters despise art-critics, and physiologists, physicists, or mathematicians have usually similar feelings; there is no scorn more profound, or on the whole more justifiable, than that of the men who make for the men who explain. Exposition, criticism, appreciation, is work for second-rate minds.

I can remember arguing this point once in one of the few serious conversations that I ever had with Housman. Housman, in his Leslie Stephen lecture *The Name and Nature of Poetry*, had denied very emphatically that he was a 'critic'; but he had denied it in what seemed to me a singularly perverse way, and had expressed an admiration for literary criticism which startled and scandalized me.

He had begun with a quotation from his inaugural lecture, delivered twenty-two years before—

> Whether the faculty of literary criticism is the best gift that Heaven has in its treasuries, I cannot say; but Heaven seems to think so, for assuredly it is the gift most charily bestowed. Orators and poets..., if rare in comparison with blackberries, are commoner than returns of Halley's comet: literary critics are less common....

And he had continued—

> In these twenty-two years I have improved in some respects and deteriorated in others, but I have not so much improved as to become a literary critic, nor so much deteriorated as to fancy that I have become one.

It had seemed to me deplorable that a great scholar and a fine poet should write like this, and, finding myself next to him in Hall a few weeks later, I plunged in and said so. Did he really mean what he had said to be taken very seriously? Would the life of the best of critics really have seemed to him comparable with that of a scholar and a poet? We argued these questions all through dinner, and I think that finally he agreed with me. I must not seem to claim a dialectical triumph over a man who can no longer contradict me; but 'Perhaps not entirely' was, in the end, his reply to the first question, and 'Probably no' to the second.

There may have been some doubt about Housman's feelings, and I do not wish to claim him as on my side; but there is no doubt at all about the feelings of men of science, and I share them fully. If then I find myself writing, not mathematics but 'about' mathematics, it is a confession of weakness, for which I may rightly be scorned or pitied by younger and more vigorous mathematicians. I write about mathematics because, like any other mathematician who has passed sixty, I have no longer the freshness of mind, the energy, or the patience to carry on effectively with my proper job.

2

I PROPOSE to put forward an apology for mathematics; and I may be told that it needs none, since there are now few studies more generally recognized, for good reasons or bad, as profitable and praiseworthy. This may be true; indeed it is probable, since the sensational triumphs of Einstein, that stellar astronomy and atomic physics are the only sciences which stand higher in popular estimation. A mathematician need not now consider himself on the defensive. He does not have to meet the sort of opposition described by Bradley in the admirable defence of metaphysics which forms the introduction to *Appearance and Reality*.

A metaphysician, says Bradley, will be told that 'metaphysical knowledge is wholly impossible', or that 'even if possible to a certain degree, it is practically no knowledge worth the name'. 'The same problems,' he will hear, 'the same disputes, the same sheer failure. Why not abandon it and come out? Is there nothing else more worth your labour?' There is no one so stupid as to use this sort of language about mathematics. The mass of mathematical truth is obvious and imposing; its practical applications, the bridges and steam-engines and dynamos, obtrude themselves on the dullest imagination. The public does not need to be convinced that there is something in mathematics.

All this is in its way very comforting to mathematicians, but it is hardly possible for a genuine mathematician to be content with it. Any genuine mathematician must feel that it is not on these crude achievements that the real case for mathematics rests, that the popular reputation of mathematics is based largely on ignorance and confusion, and that there is room for a more rational defence. At any rate, I am disposed to try to make one. It should be a simpler task than Bradley's difficult apology.

I shall ask, then, why is it really worth while to make a serious study of mathematics? What is the proper justification of a mathematician's life? And my answers will be, for the most part, such as are to be expected from a mathematician: I think that it is worth while, that there is ample justification. But I should say at once that my defence of mathematics will be a defence of myself, and that my apology is bound to be to some extent egotistical. I should not think it worth while to apologize for my subject if I regarded myself as one of its failures.

Some egotism of this sort is inevitable, and I do not feel that it really needs justification. Good work is not done by 'humble' men. It is one of the first duties of a professor, for example, in any subject, to exaggerate a little both the importance of his subject and his own importance in it. A man who is always asking 'Is what I do worth while?' and 'Am I the right person to do it?' will always be ineffective himself and a discouragement to others. He must shut his eyes a little and think a little more of his subject and himself than they deserve. This is not too difficult: it is harder not to make his subject and himself ridiculous by shutting his eyes too tightly.

3

A MAN who sets out to justify his existence and his activities has to distinguish two different questions. The first is whether the work which he does is worth doing; and the second is why he does it, whatever its value may be. The first question is often very difficult, and the answer very discouraging, but most people will find the second easy enough even then. Their answers, if they are honest, will usually take one or other of two forms; and the second form is merely a humbler variation of the first, which is the only answer which we need consider seriously.

(1) 'I do what I do because it is the one and only thing that I can do at all well. I am a lawyer, or a stockbroker, or a professional cricketer, because I have some real talent for that particular job. I am a lawyer because I have a fluent tongue, and am interested in legal subtleties; I am a stockbroker because my judgement of the markets is quick and sound; I am a professional cricketer because I can bat unusually well. I agree that it might be better to be a poet or a mathematician, but unfortunately I have no talent for such pursuits.'

I am not suggesting that this is a defence which can be made by most people, since most people can do nothing at all well. But it is impregnable when it can be made without absurdity, as it can by a substantial minority: perhaps five or even ten per cent of men can do something rather well. It is a tiny minority who can do anything *really* well, and the number of men who can do two things well is negligible. If a man has any genuine talent, he should be ready to make almost any sacrifice in order to cultivate it to the full.

This view was endorsed by Dr Johnson—

> When I told him that I had been to see [his namesake] Johnson ride upon three horses, he said 'Such a man, sir, should be encouraged, for his performances show the extent of the human powers...'—

and similarly he would have applauded mountain climbers, channel swimmers, and blindfold chess-players. For my own part, I am entirely in sympathy with all such attempts at remarkable achievement. I feel some sympathy even with conjurors and ventriloquists; and when Alekhine and Bradman set out to beat records, I am quite bitterly disappointed if they fail. And here both Dr Johnson and I find ourselves in agreement with the public. As W. J. Turner has said so truly, it is only the ‘highbrows’ (in the unpleasant sense) who do not admire the ‘real swells’.

We have of course to take account of the differences in value between different activities. I would rather be a novelist or a painter than a statesman of similar rank; and there are many roads to fame which most of us would reject as actively pernicious. Yet it is seldom that such differences of value will turn the scale in a man’s choice of a career, which will almost always be dictated by the limitations of his natural abilities. Poetry is more valuable than cricket, but Bradman would be a fool if he sacrificed his cricket in order to write second-rate minor poetry (and I suppose that it is unlikely that he could do better). If the cricket were a little less supreme, and the poetry better, then the choice might be more difficult: I do not know whether I would rather have been Victor Trumper or Rupert Brooke. It is fortunate that such dilemmas occur so seldom.

I may add that they are particularly unlikely to present themselves to a mathematician. It is usual to exaggerate rather grossly the differences between the mental processes of mathematicians and other people, but it is undeniable that a gift for mathematics is one of the most specialized talents, and that mathematicians as a class are not particularly distinguished for general ability or versatility. If a man is in any sense a real mathematician, then it is a hundred to one that his mathematics will be far better than anything else he can do, and that he would be silly if he surrendered any decent opportunity of exercising his one talent in order to do undistinguished work in other fields. Such a sacrifice could be justified only by economic necessity or age.

4

I HAD better say something here about this question of age, since it is particularly important for mathematicians. No mathematician should ever allow himself to forget that mathematics, more than any other art or science, is a young man's game. To take a simple illustration at a comparatively humble level, the average age of election to the Royal Society is lowest in mathematics.

We can naturally find much more striking illustrations. We may consider, for example, the career of a man who was certainly one of the world's three greatest mathematicians. Newton gave up mathematics at fifty, and had lost his enthusiasm long before; he had recognized no doubt by the time that he was forty that his great creative days were over. His greatest ideas of all, fluxions and the law of gravitation, came to him about 1666, when he was twenty-four—'in those days I was in the prime of my age for invention, and minded mathematics and philosophy more than at any time since'. He made big discoveries until he was nearly forty (the 'elliptic orbit' at thirty-seven), but after that he did little but polish and perfect.

Galois died at twenty-one, Abel at twenty-seven, Ramanujan at thirty-three, Riemann at forty. There have been men who have done great work a good deal later; Gauss's great memoir on differential geometry was published when he was fifty (though he had had the fundamental ideas ten years before). I do not know an instance of a major mathematical advance initiated by a man past fifty. If a man of mature age loses interest in and abandons mathematics, the loss is not likely to be very serious either for mathematics or for himself.

On the other hand the gain is no more likely to be substantial; the later records of mathematicians who have left mathematics are not particularly encouraging. Newton made a quite competent Master of the Mint (when he was not quarrelling with anybody). Painlevé was a not very successful Premier of France. Laplace's political career was highly discreditable, but he is hardly a fair instance, since he was dishonest rather than incompetent, and never really 'gave up' mathematics. It is very hard to find an instance of a first-rate mathematician who has abandoned mathematics and attained first-rate distinction in any other field.* There may have been young men who would have been first-rate mathematicians if they had stuck to mathematics, but I have never heard of a

* Pascal seems the best.

really plausible example. And all this is fully borne out by my own very limited experience. Every young mathematician of real talent whom I have known has been faithful to mathematics, and not from lack of ambition but from abundance of it; they have all recognized that there, if anywhere, lay the road to a life of any distinction.

5

THERE is also what I called the 'humbler variation' of the standard apology; but I may dismiss this in a very few words.

(2) 'There is *nothing* that I can do particularly well. I do what I do because it came my way. I really never had a chance of doing anything else.' And this apology too I accept as conclusive. It is quite true that most people can do nothing well. If so, it matters very little what career they choose, and there is really nothing more to say about it. It is a conclusive reply, but hardly one likely to be made by a man with any pride; and I may assume that none of us would be content with it.

6

It is time to begin thinking about the first question which I put in § 3, and which is so much more difficult than the second. Is mathematics, what I and other mathematicians mean by mathematics, worth doing; and if so, why?

I have been looking again at the first pages of the inaugural lecture which I gave at Oxford in 1920, where there is an outline of an apology for mathematics. It is very inadequate (less than a couple of pages), and it is written in a style (a first essay, I suppose, in what I then imagined to be the 'Oxford manner') of which I am not now particularly proud; but I still feel that, however much development it may need, it contains the essentials of the matter. I will resume what I said then, as a preface to a fuller discussion.

(1) I began by laying stress on the *harmlessness* of mathematics—'the study of mathematics is, if an unprofitable, a perfectly harmless and innocent occupation'. I shall

stick to that, but obviously it will need a good deal of expansion and explanation.

Is mathematics 'unprofitable'? In some ways, plainly, it is not; for example, it gives great pleasure to quite a large number of people. I was thinking of 'profit', however, in a narrower sense. Is mathematics 'useful', *directly* useful, as other sciences such as chemistry and physiology are? This is not an altogether easy or uncontroversial question, and I shall ultimately say No, though some mathematicians, and most outsiders, would no doubt say Yes. And is mathematics 'harmless'? Again the answer is not obvious, and the question is one which I should have in some ways preferred to avoid, since it raises the whole problem of the effect of science on war. Is mathematics harmless, in the sense in which, for example, chemistry plainly is not? I shall have to come back to both these questions later.

(2) I went on to say that 'the scale of the universe is large and, if we are wasting our time, the waste of the lives of a few university dons is no such overwhelming catastrophe':

and here I may seem to be adopting, or affecting, the pose of exaggerated humility which I repudiated a moment ago. I am sure that that was not what was really in my mind; I was trying to say in a sentence what I have said at much greater length in § 3. I was assuming that we dons really had our little talents, and that we could hardly be wrong if we did our best to cultivate them fully.

(3) Finally (in what seem to me now some rather painfully rhetorical sentences) I emphasized the permanence of mathematical achievement—

> What we do may be small, but it has a certain character of permanence; and to have produced anything of the slightest permanent interest, whether it be a copy of verses or a geometrical theorem, is to have done something utterly beyond the powers of the vast majority of men.

And—

> In these days of conflict between ancient and modern studies, there must surely be something to be said for a study which did not begin with Pythagoras, and will not end with Einstein, but is the oldest and the youngest of all.

All this is 'rhetoric'; but the substance of it seems to me still to ring true, and I can expand it at once without prejudging any of the other questions which I am leaving open.

7

I SHALL assume that I am writing for readers who are full, or have in the past been full, of a proper spirit of ambition. A man's first duty, a young man's at any rate, is to be ambitious. Ambition is a noble passion which may legitimately take many forms; there was *something* noble in the ambition of Attila or Napoleon: but the noblest ambition is that of leaving behind one something of permanent value—

> Here, on the level sand,
> Between the sea and land,
> What shall I build or write
> Against the fall of night?
>
> Tell me of runes to grave
> That hold the bursting wave,
> Or bastions to design
> For longer date than mine.

Ambition has been the driving force behind nearly all the best work of the world. In particular, practically all substantial contributions to human happiness have been made by ambitious men. To take two famous examples, were not Lister and Pasteur ambitious? Or, on a humbler level, King Gillette and William Willett; and who in recent times have contributed more to human comfort than they?

Physiology provides particularly good examples, just because it is so obviously a 'beneficial' study. We must guard against a fallacy common among apologists of science. the fallacy of supposing that the men whose work most benefits humanity are thinking much of that while they do it, that physiologists, for example, have particularly noble souls. A physiologist may indeed be glad to remember that his work will benefit mankind, but the motives which provide the force and the inspiration for it are indistinguishable from those of a classical scholar or a mathematician.

There are many highly respectable motives

which may lead men to prosecute research, but three which are much more important than the rest. The first (without which the rest must come to nothing) is intellectual curiosity, desire to know the truth. Then, professional pride, anxiety to be satisfied with one's performance, the shame that overcomes any self-respecting craftsman when his work is unworthy of his talent. Finally, ambition, desire for reputation, and the position, even the power or the money, which it brings. It may be fine to feel, when you have done your work, that you have added to the happiness or alleviated the sufferings of others, but that will not be why you did it. So if a mathematician, or a chemist, or even a physiologist, were to tell me that the driving force in his work had been the desire to benefit humanity, then I should not believe him (nor should I think the better of him if I did). His dominant motives have been those which I have stated, and in which, surely, there is nothing of which any decent man need be ashamed.

8

IF intellectual curiosity, professional pride, and ambition are the dominant incentives to research, then assuredly no one has a fairer chance of gratifying them than a mathematician. His subject is the most curious of all—there is none in which truth plays such odd pranks. It has the most elaborate and the most fascinating technique, and gives unrivalled openings for the display of sheer professional skill. Finally, as history proves abundantly, mathematical achievement, whatever its intrinsic worth, is the most enduring of all.

We can see this even in semi-historic civilizations. The Babylonian and Assyrian civilizations have perished; Hammurabi, Sargon, and Nebuchadnezzar are empty names; yet Babylonian mathematics is still interesting, and the Babylonian scale of 60 is still used in astronomy. But of course the crucial case is that of the Greeks.

The Greeks were the first mathematicians

who are still 'real' to us to-day. Oriental mathematics may be an interesting curiosity, but Greek mathematics is the real thing. The Greeks first spoke a language which modern mathematicians can understand; as Littlewood said to me once, they are not clever schoolboys or 'scholarship candidates', but 'Fellows of another college'. So Greek mathematics is 'permanent', more permanent even than Greek literature. Archimedes will be remembered when Aeschylus is forgotten, because languages die and mathematical ideas do not. 'Immortality' may be a silly word, but probably a mathematician has the best chance of whatever it may mean.

Nor need he fear very seriously that the future will be unjust to him. Immortality is often ridiculous or cruel: few of us would have chosen to be Og or Ananias or Gallio. Even in mathematics, history sometimes plays strange tricks; Rolle figures in the text-books of elementary calculus as if he had been a mathematician like Newton; Farey is immortal because he failed to understand a theorem which Haros had proved perfectly fourteen years before; the names of five worthy Norwegians still stand in Abel's *Life*, just for one act of conscientious imbecility, dutifully performed at the expense of their country's greatest man. But on the whole the history of science is fair, and this is particularly true in mathematics. No other subject has such clear-cut or unanimously accepted standards, and the men who are remembered are almost always the men who merit it. Mathematical fame, if you have the cash to pay for it, is one of the soundest and steadiest of investments.

9

ALL this is very comforting for dons, and especially for professors of mathematics. It is sometimes suggested, by lawyers or politicians or business men, that an academic career is one sought mainly by cautious and unambitious persons who care primarily for comfort and security. The reproach is quite misplaced. A don surrenders something, and in particular the chance of making large sums of money—it is very hard for a professor to make

£2000 a year; and security of tenure is naturally one of the considerations which make this particular surrender easy. That is not why Housman would have refused to be Lord Simon or Lord Beaverbrook. He would have rejected their careers because of his ambition, because he would have scorned to be a man to be forgotten in twenty years.

Yet how painful it is to feel that, with all these advantages, one may fail. I can remember Bertrand Russell telling me of a horrible dream. He was in the top floor of the University Library, about A.D. 2100. A library assistant was going round the shelves carrying an enormous bucket, taking down book after book, glancing at them, restoring them to the shelves or dumping them into the bucket. At last he came to three large volumes which Russell could recognize as the last surviving copy of *Principia mathematica.* He took down one of the volumes, turned over a few pages, seemed puzzled for a moment by the curious symbolism, closed the volume, balanced it in his hand and hesitated....

10

A MATHEMATICIAN, like a painter or a poet, is a maker of patterns. If his patterns are more permanent than theirs, it is because they are made with *ideas*. A painter makes patterns with shapes and colours, a poet with words. A painting may embody an 'idea', but the idea is usually commonplace and unimportant. In poetry, ideas count for a good deal more; but, as Housman insisted, the importance of ideas in poetry is habitually exaggerated: 'I cannot satisfy myself that there are any such things as poetical ideas.... Poetry is not the thing said but a way of saying it.'

> Not all the water in the rough rude sea
> Can wash the balm from an anointed King.

Could lines be better, and could ideas be at once more trite and more false? The poverty of the ideas seems hardly to affect the beauty of the verbal pattern. A mathematician, on the other hand, has no material to work with but ideas, and so his patterns are likely to last longer, since ideas wear less with time than words.

The mathematician's patterns, like the painter's or the poet's, must be *beautiful*; the ideas, like the colours or the words, must fit together in a harmonious way. Beauty is the first test: there is no permanent place in the world for ugly mathematics. And here I must deal with a misconception which is still widespread (though probably much less so now than it was twenty years ago), what Whitehead has called the 'literary superstition' that love of and aesthetic appreciation of mathematics is 'a monomania confined to a few eccentrics in each generation'.

It would be difficult now to find an educated man quite insensitive to the aesthetic appeal of mathematics. It may be very hard to *define* mathematical beauty, but that is just as true of beauty of any kind—we may not know quite what we mean by a beautiful poem, but that does not prevent us from recognizing one when we read it. Even Professor Hogben, who is out to minimize at all costs the importance of the aesthetic element in mathematics, does not venture to deny its reality. 'There are, to be sure, individuals for whom mathematics exercises a coldly impersonal attraction.... The aesthetic appeal of mathematics may be very real for a chosen few.' But they are 'few', he suggests, and they feel 'coldly' (and are really rather ridiculous people, who live in silly little university towns sheltered from the fresh breezes of the wide open spaces). In this he is merely echoing Whitehead's 'literary superstition'.

The fact is that there are few more 'popular' subjects than mathematics. Most people have some appreciation of mathematics, just as most people can enjoy a pleasant tune; and there are probably more people really interested in mathematics than in music. Appearances may suggest the contrary, but there are easy explanations. Music can be used to stimulate mass emotion, while mathematics cannot; and musical incapacity is recognized (no doubt rightly) as mildly discreditable, whereas most people are so frightened of the name of mathematics that they are ready, quite unaffectedly, to exaggerate their own mathematical stupidity.

A very little reflection is enough to expose the absurdity of the 'literary superstition'. There are masses of chess-players in every civilized country—in Russia, almost the whole educated population; and every chess-player can recognize and appreciate a 'beautiful' game or problem. Yet a chess problem is *simply* an exercise in pure mathematics (a game not entirely, since psychology also plays a part), and everyone who calls a problem 'beautiful' is applauding mathematical beauty, even if it is beauty of a comparatively lowly kind. Chess problems are the hymn-tunes of mathematics.

We may learn the same lesson, at a lower level but for a wider public, from bridge, or descending further, from the puzzle columns of the popular newspapers. Nearly all their immense popularity is a tribute to the drawing power of rudimentary mathematics, and the better makers of puzzles, such as Dudeney or 'Caliban', use very little else. They know their business; what the public wants is a little intellectual 'kick', and nothing else has quite the kick of mathematics.

I might add that there is nothing in the world which pleases even famous men (and men who have used disparaging language about mathematics) quite so much as to discover, or rediscover, a genuine mathematical theorem. Herbert Spencer republished in his autobiography a theorem about circles which he proved when he was twenty (not knowing that it had been proved over two thousand years before by Plato). Professor Soddy is a more recent and a more striking example (but *his* theorem really is his own)*.

11

A CHESS problem is genuine mathematics, but it is in some way 'trivial' mathematics. However ingenious and intricate, however original and surprising the moves, there is something essential lacking. Chess problems are *unim-*

* See his letters on the 'Hexlet' in *Nature*, vols. 137–9 (1936–7).

portant. The best mathematics is *serious* as well as beautiful—'important' if you like, but the word is very ambiguous, and 'serious' expresses what I mean much better.

I am not thinking of the 'practical' consequences of mathematics. I have to return to that point later: at present I will say only that if a chess problem is, in the crude sense, 'useless', then that is equally true of most of the best mathematics; that very little of mathematics is useful practically, and that that little is comparatively dull. The 'seriousness' of a mathematical theorem lies, not in its practical consequences, which are usually negligible, but in the *significance* of the mathematical ideas which it connects. We may say, roughly, that a mathematical idea is 'significant' if it can be connected, in a natural and illuminating way, with a large complex of other mathematical ideas. Thus a serious mathematical theorem, a theorem which connects significant ideas, is likely to lead to important advances in mathematics itself and even in other sciences. No chess problem has ever affected the general development of scientific thought; Pythagoras,

Newton, Einstein have in their times changed its whole direction.

The seriousness of a theorem, of course, does not *lie in* its consequences, which are merely the *evidence* for its seriousness. Shakespeare had an enormous influence on the development of the English language, Otway next to none, but that is not why Shakespeare was the better poet. He was the better poet because he wrote much better poetry. The inferiority of the chess problem, like that of Otway's poetry, lies not in its consequences but in its content.

There is one more point which I shall dismiss very shortly, not because it is uninteresting but because it is difficult, and because I have no qualifications for any serious discussion in aesthetics. The beauty of a mathematical theorem *depends* a great deal on its seriousness, as even in poetry the beauty of a line may depend to some extent on the significance of the ideas which it contains. I quoted two lines of Shakespeare as an example of the sheer beauty of a verbal pattern; but

> After life's fitful fever he sleeps well

seems still more beautiful. The pattern is just as fine, and in this case the ideas have significance and the thesis is sound, so that our emotions are stirred much more deeply. The ideas do matter to the pattern, even in poetry, and much more, naturally, in mathematics; but I must not try to argue the question seriously.

12

IT will be clear by now that, if we are to have any chance of making progress, I must produce examples of 'real' mathematical theorems, theorems which every mathematician will admit to be first-rate. And here I am very heavily handicapped by the restrictions under which I am writing. On the one hand my examples must be very simple, and intelligible to a reader who has no specialized mathematical knowledge; no elaborate preliminary explanations must be needed; and a reader must be able to follow the proofs as well as the enunciations. These conditions exclude, for instance, many of the most beautiful theorems of the theory of numbers, such as Fermat's 'two square' theorem or the law of quadratic reciprocity. And on the other hand my examples should be drawn from 'pukka' mathematics, the mathematics of the working professional mathematician; and this condition excludes a good deal which it would be comparatively easy to make intelligible but which trespasses on logic and mathematical philosophy.

I can hardly do better than go back to the Greeks. I will state and prove two of the famous theorems of Greek mathematics. They are 'simple' theorems, simple both in idea and in execution, but there is no doubt at all about their being theorems of the highest class. Each is as fresh and significant as when it was discovered—two thousand years have not written a wrinkle on either of them. Finally, both the statements and the proofs can be mastered in an hour by any intelligent reader, however slender his mathematical equipment.

1. The first is Euclid's* proof of the existence of an infinity of prime numbers.

* *Elements* IX 20. The real origin of many theorems in the *Elements* is obscure, but there seems to be no particular reason for supposing that this one is not Euclid's own.

The *prime numbers* or *primes* are the numbers

(A) 2, 3, 5, 7, 11, 13, 17, 19, 23, 29, ...

which cannot be resolved into smaller factors*. Thus 37 and 317 are prime. The primes are the material out of which all numbers are built up by multiplication: thus $666 = 2 \cdot 3 \cdot 3 \cdot 37$. Every number which is not prime itself is divisible by at least one prime (usually, of course, by several). We have to prove that there are infinitely many primes, i.e. that the series (A) never comes to an end.

Let us suppose that it does, and that

$$2, 3, 5, \ldots, P$$

is the complete series (so that P is the largest prime); and let us, on this hypothesis, consider the number Q defined by the formula

$$Q = (2 \cdot 3 \cdot 5 \cdot \ldots \cdot P) + 1.$$

It is plain that Q is not divisible by any of $2, 3, 5, \ldots, P$; for it leaves the remainder 1 when divided by any one of these numbers. But, if not itself prime, it is divisible by *some* prime, and therefore there is a prime (which

* There are technical reasons for not counting 1 as a prime.

may be Q itself) greater than any of them. This contradicts our hypothesis, that there is no prime greater than P; and therefore this hypothesis is false.

The proof is by *reductio ad absurdum*, and *reductio ad absurdum*, which Euclid loved so much, is one of a mathematician's finest weapons*. It is a far finer gambit than any chess gambit: a chess player may offer the sacrifice of a pawn or even a piece, but a mathematician offers *the game*.

13

2. My second example is Pythagoras's† proof of the 'irrationality' of $\sqrt{2}$.

A 'rational number' is a fraction $\frac{a}{b}$, where a and b are integers; we may suppose that a and b have no common factor, since if they had we could remove it. To say that '$\sqrt{2}$ is irrational' is merely another way of saying

* The proof can be arranged so as to avoid a *reductio*, and logicians of some schools would prefer that it should be.

† The proof traditionally ascribed to Pythagoras, and certainly a product of his school. The theorem occurs, in a much more general form, in Euclid (*Elements* x 9).

that 2 cannot be expressed in the form $\left(\frac{a}{b}\right)^2$;
and this is the same thing as saying that the equation

$$\text{(B)} \qquad a^2 = 2b^2$$

cannot be satisfied by integral values of a and b which have no common factor. This is a theorem of pure arithmetic, which does not demand any knowledge of 'irrational numbers' or depend on any theory about their nature.

We argue again by *reductio ad absurdum*; we suppose that (B) is true, a and b being integers without any common factor. It follows from (B) that a^2 is even (since $2b^2$ is divisible by 2), and therefore that a is even (since the square of an odd number is odd). If a is even then

$$\text{(C)} \qquad a = 2c$$

for some integral value of c; and therefore

$$2b^2 = a^2 = (2c)^2 = 4c^2$$

or

$$\text{(D)} \qquad b^2 = 2c^2.$$

Hence b^2 is even, and therefore (for the same reason as before) b is even. That is to say,

a and b are both even, and so have the common factor 2. This contradicts our hypothesis, and therefore the hypothesis is false.

It follows from Pythagoras's theorem that the diagonal of a square is incommensurable with the side (that their ratio is not a rational number, that there is no unit of which both are integral multiples). For if we take the side as our unit of length, and the length of the diagonal is d, then, by a very familiar theorem also ascribed to Pythagoras*,

$$d^2 = 1^2 + 1^2 = 2,$$

so that d cannot be a rational number.

I could quote any number of fine theorems from the theory of numbers whose *meaning* anyone can understand. For example, there is what is called 'the fundamental theorem of arithmetic', that any integer can be resolved, *in one way only*, into a product of primes. Thus $666 = 2 \cdot 3 \cdot 3 \cdot 37$, and there is no other decomposition; it is impossible that $666 = 2 \cdot 11 \cdot 29$ or that $13 \cdot 89 = 17 \cdot 73$ (and we can see so without working out the products). This

* Euclid, *Elements* I 47.

theorem is, as its name implies, the foundation of higher arithmetic; but the proof, although not 'difficult', requires a certain amount of preface and might be found tedious by an unmathematical reader.

Another famous and beautiful theorem is Fermat's 'two square' theorem. The primes may (if we ignore the special prime 2) be arranged in two classes; the primes

$$5, 13, 17, 29, 37, 41, \ldots$$

which leave remainder 1 when divided by 4, and the primes

$$3, 7, 11, 19, 23, 31, \ldots$$

which leave remainder 3. All the primes of the first class, and none of the second, can be expressed as the sum of two integral squares: thus

$$5 = 1^2+2^2, \quad 13 = 2^2+3^2,$$
$$17 = 1^2+4^2, \quad 29 = 2^2+5^2;$$

but 3, 7, 11, and 19 are not expressible in this way (as the reader may check by trial). This is Fermat's theorem, which is ranked, very justly, as one of the finest of arithmetic. Unfortunately there is no proof within the comprehension of anybody but a fairly expert mathematician.

There are also beautiful theorems in the 'theory of aggregates' (*Mengenlehre*), such as Cantor's theorem of the 'non-enumerability' of the continuum. Here there is just the opposite difficulty. The proof is easy enough, when once the language has been mastered, but considerable explanation is necessary before the *meaning* of the theorem becomes clear. So I will not try to give more examples. Those which I have given are test cases, and a reader who cannot appreciate them is unlikely to appreciate anything in mathematics.

I said that a mathematician was a maker of patterns of ideas, and that beauty and seriousness were the criteria by which his patterns should be judged. I can hardly believe that anyone who has understood the two theorems will dispute that they pass these tests. If we compare them with Dudeney's most ingenious puzzles, or the finest chess problems that masters of that art have composed, their superiority in both respects stands out: there is an unmistakable difference of class. They are much more serious, and also much more beautiful; can we define, a little more closely, where their superiority lies?

14

IN the first place, the superiority of the mathematical theorems in *seriousness* is obvious and overwhelming. The chess problem is the product of an ingenious but very limited complex of ideas, which do not differ from one another very fundamentally and have no external repercussions. We should think in the same way if chess had never been invented, whereas the theorems of Euclid and Pythagoras have influenced thought profoundly, even outside mathematics.

Thus Euclid's theorem is vital for the whole structure of arithmetic. The primes are the raw material out of which we have to build arithmetic, and Euclid's theorem assures us that we have plenty of material for the task. But the theorem of Pythagoras has wider applications and provides a better text.

We should observe first that Pythagoras's

argument is capable of far-reaching extension, and can be applied, with little change of principle, to very wide classes of 'irrationals'. We can prove very similarly (as Theodorus seems to have done) that

$$\sqrt{3}, \sqrt{5}, \sqrt{7}, \sqrt{11}, \sqrt{13}, \sqrt{17}$$

are irrational, or (going beyond Theodorus) that $\sqrt[3]{2}$ and $\sqrt[3]{17}$ are irrational*.

Euclid's theorem tells us that we have a good supply of material for the construction of a coherent arithmetic of the integers. Pythagoras's theorem and its extensions tell us that, when we have constructed this arithmetic, it will not prove sufficient for our needs, since there will be many magnitudes which obtrude themselves upon our attention and which it will be unable to measure; the diagonal of the square is merely the most obvious example. The profound importance of this discovery was recognized at once by the Greek mathematicians. They had begun by assuming (in

* See Ch. IV of Hardy and Wright's *Introduction to the Theory of Numbers*, where there are discussions of different generalizations of Pythagoras's argument, and of a historical puzzle about Theodorus.

accordance, I suppose, with the 'natural' dictates of 'common sense') that all magnitudes of the same kind are commensurable, that any two lengths, for example, are multiples of some common unit, and they had constructed a theory of proportion based on this assumption. Pythagoras's discovery exposed the unsoundness of this foundation, and led to the construction of the much more profound theory of Eudoxus which is set out in the fifth book of the *Elements*, and which is regarded by many modern mathematicians as the finest achievement of Greek mathematics. This theory is astonishingly modern in spirit, and may be regarded as the beginning of the modern theory of irrational number, which has revolutionized mathematical analysis and had much influence on recent philosophy.

There is no doubt at all, then, of the 'seriousness' of either theorem. It is therefore the better worth remarking that neither theorem has the slightest 'practical' importance. In practical applications we are concerned only with comparatively small numbers; only stellar astronomy and atomic physics deal with 'large' numbers, and they have very little more practical importance, as yet, than the most abstract pure mathematics. I do not know what is the highest degree of accuracy which is ever useful to an engineer—we shall be very generous if we say ten significant figures. Then

$$3{\cdot}14159265$$

(the value of π to eight places of decimals) is the ratio

$$\frac{314159265}{100000000}$$

of two numbers of nine digits. The number of primes less than 1,000,000,000 is 50,847,478: that is enough for an engineer, and he can be perfectly happy without the rest. So much for Euclid's theorem; and, as regards Pythagoras's, it is obvious that irrationals are uninteresting to an engineer, since he is concerned only with approximations, and all approximations are rational.

15

A 'SERIOUS' theorem is a theorem which contains 'significant' ideas, and I suppose that I ought to try to analyse a little more closely the qualities which make a mathematical idea significant. This is very difficult, and it is unlikely that any analysis which I can give will be very valuable. We can recognize a 'significant' idea when we see it, as we can those which occur in my two standard theorems; but this power of recognition requires a rather high degree of mathematical sophistication, and of that familiarity with mathematical ideas which comes only from many years spent in their company. So I must attempt some sort of analysis; and it should be possible to make one which, however inadequate, is sound and intelligible so far as it goes. There are two things at any rate which seem essential, a certain *generality* and a certain *depth*; but neither quality is easy to define at all precisely.

A significant mathematical idea, a serious mathematical theorem, should be 'general' in some such sense as this. The idea should be one which is a constituent in many mathematical constructs, which is used in the proof of theorems of many different kinds. The theorem should be one which, even if stated originally (like Pythagoras's theorem) in a quite special form, is capable of considerable extension and is typical of a whole class of theorems of its kind. The relations revealed by the proof should be such as connect many different mathematical ideas. All this is very vague, and subject to many reservations. But it is easy enough to see that a theorem is unlikely to be serious when it lacks these qualities conspicuously; we have only to take examples from the isolated curiosities in which arithmetic abounds. I take two, almost at random, from Rouse Ball's *Mathematical Recreations**.

(*a*) 8712 and 9801 are the only four-figure numbers which are integral multiples of their 'reversals':

$$8712 = 4 \cdot 2178, \quad 9801 = 9 \cdot 1089,$$

* 11th edition, 1939 (revised by H. S. M. Coxeter).

and there are no other numbers below 10,000 which have this property.

(*b*) There are just four numbers (after 1) which are the sums of the cubes of their digits, viz.

$$153 = 1^3+5^3+3^3, \quad 370 = 3^3+7^3+0^3,$$
$$371 = 3^3+7^3+1^3, \quad 407 = 4^3+0^3+7^3.$$

These are odd facts, very suitable for puzzle columns and likely to amuse amateurs, but there is nothing in them which appeals much to a mathematician. The proofs are neither difficult nor interesting—merely a little tiresome. The theorems are not serious; and it is plain that one reason (though perhaps not the most important) is the extreme speciality of both the enunciations and the proofs, which are not capable of any significant generalization.

16

'GENERALITY' is an ambiguous and rather dangerous word, and we must be careful not to allow it to dominate our discussion too much. It is used in various senses both in mathematics and in writings about mathematics, and there is one of these in particular, on which logicians have very properly laid great stress, which is entirely irrelevant here. In this sense, which is quite easy to define, *all* mathematical theorems are equally and completely 'general'.

'The certainty of mathematics', says Whitehead*, 'depends on its complete abstract generality.' When we assert that $2+3=5$, we are asserting a relation between three groups of 'things'; and these 'things' are not apples or pennies, or things of any one particular sort or another, but *just* things, 'any old things'. The meaning of the statement is entirely independent of the individualities of the members of the groups. All mathematical 'objects' or 'entities' or 'relations', such as '2', '3', '5', '+', or '=', and all mathematical propositions in which they occur, are completely general in the sense of being completely abstract. Indeed one of Whitehead's words is superfluous, since generality, in this sense, *is* abstractness.

* *Science and the Modern World*, p. 33.

This sense of the word is important, and the logicians are quite right to stress it, since it embodies a truism which a good many people who ought to know better are apt to forget. It is quite common, for example, for an astronomer or a physicist to claim that he has found a 'mathematical proof' that the physical universe must behave in a particular way. All such claims, if interpreted literally, are strictly nonsense. It *cannot* be possible to prove mathematically that there will be an eclipse to-morrow, because eclipses, and other physical phenomena, do not form part of the abstract world of mathematics; and this, I suppose, all astronomers would admit when pressed, however many eclipses they may have predicted correctly.

It is obvious that we are not concerned with this sort of 'generality' now. We are looking for *differences* of generality between one mathematical theorem and another, and in Whitehead's sense all are equally general. Thus the 'trivial' theorems (*a*) and (*b*) of § 15 are just as 'abstract' or 'general' as those of Euclid and Pythagoras, and so is a chess problem. It makes no difference to a chess problem whether the pieces are white and black, or red and green, or whether there are physical 'pieces' at all; it is the *same* problem which an expert carries easily in his head and which we have to reconstruct laboriously with the aid of the board. The board and the pieces are mere devices to stimulate our sluggish imaginations, and are no more essential to the problem than the blackboard and the chalk are to the theorems in a mathematical lecture.

It is not this kind of generality, common to all mathematical theorems, which we are looking for now, but the more subtle and elusive kind of generality which I tried to describe in rough terms in § 15. And we must be careful not to lay *too* much stress even on generality of this kind (as I think logicians like Whitehead tend to do). It is not mere 'piling of subtlety of generalization upon subtlety of generalization'* which is the outstanding achievement of modern mathematics. Some measure of generality must be present in any high-class theorem, but *too much* tends in-

* *Science and the Modern World*, p. 44.

evitably to insipidity. 'Everything is what it is, and not another thing', and the differences between things are quite as interesting as their resemblances. We do not choose our friends because they embody all the pleasant qualities of humanity, but because they are the people that they are. And so in mathematics; a property common to too many objects can hardly be very exciting, and mathematical ideas also become dim unless they have plenty of individuality. Here at any rate I can quote Whitehead on my side: 'it is the large generalization, limited by a happy particularity, which is the fruitful conception*.'

17

THE second quality which I demanded in a significant idea was *depth*, and this is still more difficult to define. It has *something* to do with *difficulty*; the 'deeper' ideas are usually the harder to grasp: but it is not at all the same. The ideas underlying Pythagoras's theorem and its generalizations are quite deep, but no

* *Science and the Modern World*, p. 46.

mathematician now would find them difficult. On the other hand a theorem may be essentially superficial and yet quite difficult to prove (as are many 'Diophantine' theorems, i.e. theorems about the solution of equations in integers).

It seems that mathematical ideas are arranged somehow in strata, the ideas in each stratum being linked by a complex of relations both among themselves and with those above and below. The lower the stratum, the deeper (and in general the more difficult) the idea. Thus the idea of an 'irrational' is deeper than that of an integer; and Pythagoras's theorem is, for that reason, deeper than Euclid's.

Let us concentrate our attention on the relations between the integers, or some other group of objects lying in some particular stratum. Then it may happen that one of these relations can be comprehended completely, that we can recognize and prove, for example, some property of the integers, without any knowledge of the contents of lower strata. Thus we proved Euclid's theorem by consideration of properties of integers only. But there are also many theorems about integers which we cannot appreciate properly, and still less prove, without digging deeper and considering what happens below.

It is easy to find examples in the theory of prime numbers. Euclid's theorem is very important, but not very deep: we can prove that there are infinitely many primes without using any notion deeper than that of 'divisibility'. But new questions suggest themselves as soon as we know the answer to this one. There is an infinity of primes, but how is this infinity distributed? Given a large number N, say 10^{80} or $10^{10^{10}}$,* about how many primes are there less than N?† When we ask *these* questions, we find ourselves in a quite different position. We can answer them, with rather surprising accuracy, but only by boring much deeper, leaving the integers above us for a while, and using the most powerful weapons of the modern theory of functions. Thus the

* It is supposed that the number of protons in the universe is about 10^{80}. The number $10^{10^{10}}$, if written at length, would occupy about 50,000 volumes of average size.

† As I mentioned in § 14, there are 50,847,478 primes less than 1,000,000,000; but that is as far as our *exact* knowledge extends.

theorem which answers our questions (the so-called 'Prime Number Theorem') is a much deeper theorem than Euclid's or even Pythagoras's.

I could multiply examples, but this notion of 'depth' is an elusive one even for a mathematician who can recognize it, and I can hardly suppose that I could say anything more about it here which would be of much help to other readers.

18

THERE is still one point remaining over from § 11, where I started the comparison between 'real mathematics' and chess. We may take it for granted now that in substance, seriousness, significance, the advantage of the real mathematical theorem is overwhelming. It is almost equally obvious, to a trained intelligence, that it has a great advantage in beauty also; but this advantage is much harder to define or locate, since the *main* defect of the chess problem is plainly its 'triviality', and the contrast in this respect mingles with and disturbs any more purely aesthetic judgement. What 'purely aesthetic' qualities can we distinguish in such theorems as Euclid's and Pythagoras's? I will not risk more than a few disjointed remarks.

In both theorems (and in the theorems, of course, I include the proofs) there is a very high degree of *unexpectedness*, combined with *inevitability* and *economy*. The arguments take so odd and surprising a form; the weapons used seem so childishly simple when compared with the far-reaching results; but there is no escape from the conclusions. There are no complications of detail—one line of attack is enough in each case; and this is true too of the proofs of many much more difficult theorems, the full appreciation of which demands quite a high degree of technical proficiency. We do not want many 'variations' in the proof of a mathematical theorem: 'enumeration of cases', indeed, is one of the duller forms of mathematical argument. A mathematical proof should resemble a simple and clear-cut constellation, not a scattered cluster in the Milky Way.

A chess problem also has unexpectedness, and a certain economy; it is essential that the moves should be surprising, and that every piece on the board should play its part. But the aesthetic effect is cumulative. It is essential also (unless the problem is too simple to be really amusing) that the key-move should be followed by a good many variations, each requiring its own individual answer. 'If P–B5 then Kt–R6; if .... then .... ; if .... then ....'—the effect would be spoilt if there were not a good many different replies. All this is quite genuine mathematics, and has its merits; but it is just that 'proof by enumeration of cases' (and of cases which do not, at bottom, differ at all profoundly*) which a real mathematician tends to despise.

I am inclined to think that I could reinforce my argument by appealing to the feelings of chess-players themselves. Surely a chess master, a player of great games and great matches, at bottom scorns a problemist's purely mathematical art. He has much of it

* I believe that it is now regarded as a *merit* in a problem that there should be many variations of the same type.

in reserve himself, and can produce it in an emergency: 'if he had made such and such a move, then I had such and such a winning combination in mind.' But the 'great game' of chess is primarily psychological, a conflict between one trained intelligence and another, and not a mere collection of small mathematical theorems.

19

I MUST return to my Oxford apology, and examine a little more carefully some of the points which I postponed in § 6. It will be obvious by now that I am interested in mathematics only as a creative art. But there are other questions to be considered, and in particular that of the 'utility' (or uselessness) of mathematics, about which there is much confusion of thought. We must also consider whether mathematics is really quite so 'harmless' as I took for granted in my Oxford lecture.

A science or an art may be said to be 'useful' if its development increases, even indirectly, the material well-being and comfort of men, if it promotes happiness, using that word in a crude and commonplace way. Thus medicine and physiology are useful because they relieve suffering, and engineering is useful because it helps us to build houses and bridges, and so to raise the standard of life (engineering, of course, does harm as well, but that is not the question at the moment). Now some mathematics is certainly useful in this way; the engineers could not do their job without a fair working knowledge of mathematics, and mathematics is beginning to find applications even in physiology. So here we have a possible ground for a defence of mathematics; it may not be the best, or even a particularly strong defence, but it is one which we must examine. The 'nobler' uses of mathematics, if such they be, the uses which it shares with all creative art, will be irrelevant to our examination. Mathematics may, like poetry or music, 'promote and sustain a lofty habit of mind', and so increase the happiness of mathematicians and even of other people; but to defend it on that ground would be merely to elaborate what I have said already. What we have to consider now is the 'crude' utility of mathematics.

20

ALL this may seem very obvious, but even here there is often a good deal of confusion, since the most 'useful' subjects are quite commonly just those which it is most useless for most of us to learn. It is useful to have an adequate supply of physiologists and engineers; but physiology and engineering are not useful studies for ordinary men (though their study may of course be defended on other grounds). For my own part I have never once found myself in a position where such scientific knowledge as I possess, outside pure mathematics, has brought me the slightest advantage.

It is indeed rather astonishing how little practical value scientific knowledge has for ordinary men, how dull and commonplace such of it as has value is, and how its value seems almost to vary inversely to its reputed utility. It is useful to be tolerably quick at common arithmetic (and that, of course, is pure mathematics). It is useful to know a little French or German, a little history and geography, perhaps even a little economics. But a little chemistry, physics, or physiology has no value at all in ordinary life. We know that the gas will burn without knowing its constitution; when our cars break down we take them to a garage; when our stomach is out of order, we go to a doctor or a drugstore. We live either by rule of thumb or on other people's professional knowledge.

However, this is a side issue, a matter of pedagogy, interesting only to schoolmasters who have to advise parents clamouring for a 'useful' education for their sons. Of course we do not mean, when we say that physiology is useful, that most people ought to study physiology, but that the development of physiology by a handful of experts will increase the comfort of the majority. The questions which are important for us now are, how far mathematics can claim this sort of utility, what kinds of mathematics can make the strongest claims, and how far the intensive study of mathematics, as it is understood by mathematicians, can be justified on this ground alone.

21

IT will probably be plain by now to what conclusions I am coming; so I will state them at once dogmatically and then elaborate them a little. It is undeniable that a good deal of elementary mathematics—and I use the word 'elementary' in the sense in which professional mathematicians use it, in which it includes, for example, a fair working knowledge of the differential and integral calculus—has considerable practical utility. These parts of mathematics are, on the whole, rather dull; they are just the parts which have least aesthetic value. The 'real' mathematics of the 'real' mathematicians, the mathematics of Fermat and Euler and Gauss and Abel and Riemann, is almost wholly 'useless' (and this is as true of 'applied' as of 'pure' mathematics). It is not possible to justify the life of any genuine professional mathematician on the ground of the 'utility' of his work.

But here I must deal with a misconception. It is sometimes suggested that pure mathematicians glory in the uselessness of their work*, and make it a boast that it has no practical applications. The imputation is usually based on an incautious saying attributed to Gauss, to the effect that, if mathematics is the queen of the sciences, then the theory of numbers is, because of its supreme uselessness, the queen of mathematics—I have never been able to find an exact quotation. I am sure that Gauss's saying (if indeed it be his) has been rather crudely misinterpreted. If the theory of numbers could be employed for any practical and obviously honourable purpose, if it could be turned directly to the furtherance of human happiness or the relief of human suffering, as

* I have been accused of taking this view myself. I once said that 'a science is said to be useful if its development tends to accentuate the existing inequalities in the distribution of wealth, or more directly promotes the destruction of human life', and this sentence, written in 1915, has been quoted (for or against me) several times. It was of course a conscious rhetorical flourish, though one perhaps excusable at the time when it was written.

physiology and even chemistry can, then surely neither Gauss nor any other mathematician would have been so foolish as to decry or regret such applications. But science works for evil as well as for good (and particularly, of course, in time of war); and both Gauss and lesser mathematicians may be justified in rejoicing that there is one science at any rate, and that their own, whose very remoteness from ordinary human activities should keep it gentle and clean.

22

THERE is another misconception against which we must guard. It is quite natural to suppose that there is a great difference in utility between 'pure' and 'applied' mathematics. This is a delusion: there is a sharp distinction between the two kinds of mathematics, which I will explain in a moment, but it hardly affects their utility.

How do pure and applied mathematics differ from one another? This is a question which can be answered definitely and about which there is general agreement among mathematicians. There will be nothing in the least unorthodox about my answer, but it needs a little preface.

My next two sections will have a mildly philosophical flavour. The philosophy will not cut deep, or be in any way vital to my main theses; but I shall use words which are used very frequently with definite philosophical implications, and a reader might well become confused if I did not explain how I shall use them.

I have often used the adjective 'real', and as we use it commonly in conversation. I have spoken of 'real mathematics' and 'real mathematicians', as I might have spoken of 'real poetry' or 'real poets', and I shall continue to do so. But I shall also use the word 'reality', and with two different connotations.

In the first place, I shall speak of 'physical reality', and here again I shall be using the word in the ordinary sense. By physical reality I mean the material world, the world of day and night, earthquakes and eclipses, the world which physical science tries to describe.

I hardly suppose that, up to this point, any reader is likely to find trouble with my language, but now I am near to more difficult ground. For me, and I suppose for most mathematicians, there is another reality, which I will call 'mathematical reality'; and there is no sort of agreement about the nature of mathematical reality among either mathematicians or philosophers. Some hold that it is 'mental' and that in some sense we construct it, others that it is outside and independent of us. A man who could give a convincing account of mathematical reality would have solved very many of the most difficult problems of metaphysics. If he could include physical reality in his account, he would have solved them all.

I should not wish to argue any of these questions here even if I were competent to do so, but I will state my own position dogmatically in order to avoid minor misapprehensions. I believe that mathematical reality lies outside us, that our function is to discover or *observe* it, and that the theorems which we prove, and which we describe grandiloquently as our 'creations', are simply our notes of our observations. This view has been held, in one form or another, by many philosophers of high reputation from Plato onwards, and I shall use the language which is natural to a man who holds it. A reader who does not like the philosophy can alter the language: it will make very little difference to my conclusions.

23

THE contrast between pure and applied mathematics stands out most clearly, perhaps, in geometry. There is the science of pure geometry*, in which there are many geometries, projective geometry, Euclidean geometry, non-Euclidean geometry, and so forth. Each of these geometries is a *model*, a pattern of ideas, and is to be judged by the interest and beauty of its particular pattern. It is a *map* or *picture*, the joint product of many hands, a partial and imperfect copy (yet exact so far as it extends)

* We must of course, for the purposes of this discussion, count as pure geometry what mathematicians call 'analytical' geometry.

of a section of mathematical reality. But the point which is important to us now is this, that there is one thing at any rate of which pure geometries are *not* pictures, and that is the spatio-temporal reality of the physical world. It is obvious, surely, that they cannot be, since earthquakes and eclipses are not mathematical concepts.

This may sound a little paradoxical to an outsider, but it is a truism to a geometer; and I may perhaps be able to make it clearer by an illustration. Let us suppose that I am giving a lecture on some system of geometry, such as ordinary Euclidean geometry, and that I draw figures on the blackboard to stimulate the imagination of my audience, rough drawings of straight lines or circles or ellipses. It is plain, first, that the truth of the theorems which I prove is in no way affected by the quality of my drawings. Their function is merely to bring home my meaning to my hearers, and, if I can do that, there would be no gain in having them redrawn by the most skilful draughtsman. They are pedagogical illustrations, not part of the real subject-matter of the lecture.

Now let us go a stage further. The room in which I am lecturing is part of the physical world, and has itself a certain pattern. The study of that pattern, and of the general pattern of physical reality, is a science in itself, which we may call 'physical geometry'. Suppose now that a violent dynamo, or a massive gravitating body, is introduced into the room. Then the physicists tell us that the geometry of the room is changed, its whole physical pattern slightly but definitely distorted. Do the theorems which I have proved become false? Surely it would be nonsense to suppose that the proofs of them which I have given are affected in any way. It would be like supposing that a play of Shakespeare is changed when a reader spills his tea over a page. The play is independent of the pages on which it is printed, and 'pure geometries' are independent of lecture rooms, or of any other detail of the physical world.

This is the point of view of a pure mathematician. Applied mathematicians, mathematical physicists, naturally take a different view, since they are preoccupied with the physical world itself, which also has its structure or pattern. We cannot describe this pattern exactly, as we can that of a pure geometry, but we can say something significant about it. We can describe, sometimes fairly accurately, sometimes very roughly, the relations which hold between some of its constituents, and compare them with the exact relations holding between constituents of some system of pure geometry. We may be able to trace a certain resemblance between the two sets of relations, and then the pure geometry will become interesting to physicists; it will give us, to that extent, a map which 'fits the facts' of the physical world. The geometer offers to the physicist a whole set of maps from which to choose. One map, perhaps, will fit the facts better than others, and then the geometry which provides that particular map will be the geometry most important for applied mathematics. I may add that even a pure mathematician may find his appreciation of this geometry quickened, since there is no mathematician so pure that he feels no interest at all in the physical world; but, in so far as he succumbs to this temptation, he will be abandoning his purely mathematical position.

24

THERE is another remark which suggests itself here and which physicists may find paradoxical, though the paradox will probably seem a good deal less than it did eighteen years ago. I will express it in much the same words which I used in 1922 in an address to Section A of the British Association. My audience then was composed almost entirely of physicists, and I may have spoken a little provocatively on that account; but I would still stand by the substance of what I said.

I began by saying that there is probably less difference between the positions of a mathematician and of a physicist than is generally supposed, and that the most important seems to me to be this, that the mathematician is in much more direct contact with reality. This may seem a paradox, since it is the physicist who deals with the subject-matter usually described as 'real'; but a very little reflection is enough to show that the physicist's reality, whatever it may be, has few or none of the attributes which common sense ascribes instinctively to reality. A chair may be a collection of whirling electrons, or an idea in the mind of God: each of these accounts of it may have its merits, but neither conforms at all closely to the suggestions of common sense.

I went on to say that neither physicists nor philosophers have ever given any convincing account of what 'physical reality' is, or of how the physicist passes, from the confused mass of fact or sensation with which he starts, to the construction of the objects which he calls 'real'. Thus we cannot be said to know what the subject-matter of physics is; but this need not prevent us from understanding roughly what a physicist is trying to do. It is plain that he is trying to correlate the incoherent body of crude fact confronting him with some definite and orderly scheme of abstract relations, the kind of scheme which he can borrow only from mathematics.

A mathematician, on the other hand, is working with his own mathematical reality.

Of this reality, as I explained in § 22, I take a 'realistic' and not an 'idealistic' view. At any rate (and this was my main point) this realistic view is much more plausible of mathematical than of physical reality, because mathematical objects are so much more what they seem. A chair or a star is not in the least like what it seems to be; the more we think of it, the fuzzier its outlines become in the haze of sensation which surrounds it; but '2' or '317' has nothing to do with sensation, and its properties stand out the more clearly the more closely we scrutinize it. It may be that modern physics fits best into some framework of idealistic philosophy—I do not believe it, but there are eminent physicists who say so. Pure mathematics, on the other hand, seems to me a rock on which all idealism founders: 317 is a prime, not because we think so, or because our minds are shaped in one way rather than another, but *because it is so*, because mathematical reality is built that way.

25

THESE distinctions between pure and applied mathematics are important in themselves, but they have very little bearing on our discussion of the 'usefulness' of mathematics. I spoke in § 21 of the 'real' mathematics of Fermat and other great mathematicians, the mathematics which has permanent aesthetic value, as for example the best Greek mathematics has, the mathematics which is eternal because the best of it may, like the best literature, continue to cause intense emotional satisfaction to thousands of people after thousands of years. These men were all primarily pure mathematicians (though the distinction was naturally a good deal less sharp in their days than it is now); but I was not thinking only of pure mathematics. I count Maxwell and Einstein, Eddington and Dirac, among 'real' mathematicians. The great modern achievements of applied mathematics have been in relativity and quantum mechanics, and these subjects are, at present at any rate, almost as 'useless' as the theory of numbers. It is the dull and elementary parts of applied mathematics, as it is the dull and elementary parts of pure mathematics, that work for good or ill. Time may change all this. No one foresaw the applications of matrices and groups and other purely mathematical theories to modern physics, and it may be that some of the 'highbrow' applied mathematics will become 'useful' in as unexpected a way; but the evidence so far points to the conclusion that, in one subject as in the other, it is what is commonplace and dull that counts for practical life.

I can remember Eddington giving a happy example of the unattractiveness of 'useful' science. The British Association held a meeting in Leeds, and it was thought that the members might like to hear something of the applications of science to the 'heavy woollen' industry. But the lectures and demonstrations arranged for this purpose were rather a fiasco. It appeared that the members (whether citizens of Leeds or not) wanted to be entertained, and that 'heavy wool' is not at all an entertaining subject. So the attendance at these lectures was very disappointing; but those who lectured on the excavations at Knossos, or on relativity, or on the theory of prime numbers, were delighted by the audiences that they drew.

26

WHAT parts of mathematics are useful?

First, the bulk of school mathematics, arithmetic, elementary algebra, elementary Euclidean geometry, elementary differential and integral calculus. We must except a certain amount of what is taught to 'specialists', such as projective geometry. In applied mathematics, the elements of mechanics (electricity, as taught in schools, must be classified as physics).

Next, a fair proportion of university mathematics is also useful, that part of it which is really a development of school mathematics with a more finished technique, and a certain amount of the more physical subjects such as electricity and hydromechanics. We must also remember that a reserve of knowledge is always an advantage, and that the most practical of mathematicians may be seriously handicapped if his knowledge is the bare minimum which is essential to him; and for this reason we must add a little under every heading. But our general conclusion must be that such mathematics is useful as is wanted by a superior engineer or a moderate physicist; and that is roughly the same thing as to say, such mathematics as has no particular aesthetic merit. Euclidean geometry, for example, is useful in so far as it is dull—we do not want the axiomatics of parallels, or the theory of proportion, or the construction of the regular pentagon.

One rather curious conclusion emerges, that pure mathematics is on the whole distinctly more useful than applied. A pure mathematician seems to have the advantage on the practical as well as on the aesthetic side. For what is useful above all is *technique*, and mathematical technique is taught mainly through pure mathematics.

I hope that I need not say that I am not

trying to decry mathematical physics, a splendid subject with tremendous problems where the finest imaginations have run riot. But is not the position of an ordinary applied mathematician in some ways a little pathetic? If he wants to be useful, he must work in a humdrum way, and he cannot give full play to his fancy even when he wishes to rise to the heights. 'Imaginary' universes are so much more beautiful than this stupidly constructed 'real' one; and most of the finest products of an applied mathematician's fancy must be rejected, as soon as they have been created, for the brutal but sufficient reason that they do not fit the facts.

The general conclusion, surely, stands out plainly enough. If useful knowledge is, as we agreed provisionally to say, knowledge which is likely, now or in the comparatively near future, to contribute to the material comfort of mankind, so that mere intellectual satisfaction is irrelevant, then the great bulk of higher mathematics is useless. Modern geometry and algebra, the theory of numbers, the theory of aggregates and functions, relativity, quantum mechanics—no one of them stands the test much better than another, and there is no real mathematician whose life can be justified on this ground. If this be the test, then Abel, Riemann, and Poincaré wasted their lives; their contribution to human comfort was negligible, and the world would have been as happy a place without them.

27

IT may be objected that my concept of 'utility' has been too narrow, that I have defined it in terms of 'happiness' or 'comfort' only, and have ignored the general 'social' effects of mathematics on which recent writers, with very different sympathies, have laid so much stress. Thus Whitehead (who has been a mathematician) speaks of 'the tremendous effect of mathematical knowledge on the lives of men, on their daily avocations, on the organization of society'; and Hogben (who is as unsympathetic to what I and other mathematicians call mathematics as Whitehead is sympathetic) says that 'without a knowledge of mathematics, the grammar of size and order, we cannot plan the rational society in which there will be leisure for all and poverty for none' (and much more to the same effect).

I cannot really believe that all this eloquence will do much to comfort mathematicians. The language of both writers is violently exaggerated, and both of them ignore very obvious distinctions. This is very natural in Hogben's case, since he is admittedly not a mathematician; he means by 'mathematics' the mathematics which he can understand, and which I have called 'school' mathematics. *This* mathematics has many uses, which I have admitted, which we can call 'social' if we please, and which Hogben enforces with many interesting appeals to the history of mathematical discovery. It is this which gives his book its merit, since it enables him to make plain, to many readers who never have been and never will be mathematicians, that there is more in mathematics than they thought. But he has hardly any understanding of 'real' mathematics (as any one who reads what he says about Pythagoras's theorem, or about Euclid and Einstein, can tell at once), and still less sympathy with it (as he spares no pains to show). 'Real' mathematics is to him merely an object of contemptuous pity.

It is not lack of understanding or of sympathy which is the trouble in Whitehead's case; but he forgets, in his enthusiasm, distinctions with which he is quite familiar. The mathematics which has this 'tremendous effect' on the 'daily avocations of men' and on 'the organization of society' is not the Whitehead but the Hogben mathematics. The mathematics which can be used 'for ordinary purposes by ordinary men' is negligible, and that which can be used by economists or sociologists hardly rises to 'scholarship standard'. The Whitehead mathematics may affect astronomy or physics profoundly, philosophy very appreciably—high thinking of one kind is always likely to affect high thinking of another—but it has extremely little effect on anything else. Its 'tremendous effects' have been, not on men generally, but on men like Whitehead himself.

28

THERE are then two mathematics. There is the real mathematics of the real mathematicians, and there is what I will call the 'trivial' mathematics, for want of a better word. The trivial mathematics may be justified by arguments which would appeal to Hogben, or other writers of his school, but there is no such defence for the real mathematics, which must be justified as art if it can be justified at all. There is nothing in the least paradoxical or unusual in this view, which is that held commonly by mathematicians.

We have still one more question to consider. We have concluded that the trivial mathematics is, on the whole, useful, and that the real mathematics, on the whole, is not; that the trivial mathematics does, and the real mathematics does not, 'do good' in a certain sense; but we have still to ask whether either sort of mathematics does *harm*. It would be paradoxical to suggest that mathematics of any sort does much harm in time of peace, so that we are driven to the consideration of the effects of mathematics on war. It is very difficult to argue such questions at all dispassionately now, and I should have preferred to avoid them; but some sort of discussion seems inevitable. Fortunately, it need not be a long one.

There is one comforting conclusion which is easy for a real mathematician. Real mathematics has no effects on war. No one has yet discovered any warlike purpose to be served by the theory of numbers or relativity, and it seems very unlikely that anyone will do so for many years. It is true that there are branches of applied mathematics, such as ballistics and aerodynamics, which have been developed deliberately for war and demand a quite elaborate technique: it is perhaps hard to call them 'trivial', but none of them has any claim to rank as 'real'. They are indeed repulsively ugly and intolerably dull; even Littlewood could not make ballistics respectable, and if he could not who can? So a real mathematician has his conscience clear; there is nothing to be set against any value his work may have;

mathematics is, as I said at Oxford, a 'harmless and innocent' occupation.

The trivial mathematics, on the other hand, has many applications in war. The gunnery experts and aeroplane designers, for example, could not do their work without it. And the general effect of these applications is plain: mathematics facilitates (if not so obviously as physics or chemistry) modern, scientific, 'total' war.

It is not so clear as it might seem that this is to be regretted, since there are two sharply contrasted views about modern scientific war. The first and the most obvious is that the effect of science on war is merely to magnify its horror, both by increasing the sufferings of the minority who have to fight and by extending them to other classes. This is the most natural and the orthodox view. But there is a very different view which seems also quite tenable, and which has been stated with great force by Haldane in *Callinicus**. It can be maintained that modern warfare is *less* horrible than the

* J. B. S. Haldane, *Callinicus: a Defence of Chemical Warfare* (1924).

warfare of pre-scientific times; that bombs are probably more merciful than bayonets; that lachrymatory gas and mustard gas are perhaps the most humane weapons yet devised by military science; and that the orthodox view rests solely on loose-thinking sentimentalism*. It may also be urged (though this was not one of Haldane's theses) that the equalization of risks which science was expected to bring would be in the long run salutary; that a civilian's life is not worth more than a soldier's, nor a woman's than a man's; that anything is better than the concentration of savagery on one particular class; and that, in short, the sooner war comes 'all out' the better.

I do not know which of these views is nearer to the truth. It is an urgent and a moving question, but I need not argue it here. It concerns only the 'trivial' mathematics, which it would be Hogben's business to defend rather than mine. The case for his mathematics may

* I do not wish to prejudge the question by this much misused word; it may be used quite legitimately to indicate certain types of unbalanced emotion. Many people, of course, use 'sentimentalism' as a term of abuse for other people's decent feelings, and 'realism' as a disguise for their own brutality.

be rather more than a little soiled; the case for mine is unaffected.

Indeed, there is more to be said, since there is one purpose at any rate which the real mathematics may serve in war. When the world is mad, a mathematician may find in mathematics an incomparable anodyne. For mathematics is, of all the arts and sciences, the most austere and the most remote, and a mathematician should be of all men the one who can most easily take refuge where, as Bertrand Russell says, 'one at least of our nobler impulses can best escape from the dreary exile of the actual world'. It is a pity that it should be necessary to make one very serious reservation—he must not be too old. Mathematics is not a contemplative but a creative subject; no one can draw much consolation from it when he has lost the power or the desire to create; and that is apt to happen to a mathematician rather soon. It is a pity, but in that case he does not matter a great deal anyhow, and it would be silly to bother about him.

29

I WILL end with a summary of my conclusions, but putting them in a more personal way. I said at the beginning that anyone who defends his subject will find that he is defending himself; and my justification of the life of a professional mathematician is bound to be, at bottom, a justification of my own. Thus this concluding section will be in its substance a fragment of autobiography.

I cannot remember ever having wanted to be anything but a mathematician. I suppose that it was always clear that my specific abilities lay that way, and it never occurred to me to question the verdict of my elders. I do not remember having felt, as a boy, any *passion* for mathematics, and such notions as I may have had of the career of a mathematician were far from noble. I thought of mathematics in terms of examinations and scholarships: I wanted to beat other boys, and this seemed to be the way in which I could do so most decisively.

I was about fifteen when (in a rather odd way) my ambitions took a sharper turn. There is a book by 'Alan St Aubyn'* called *A Fellow of Trinity*, one of a series dealing with what is supposed to be Cambridge college life. I suppose that it is a worse book than most of Marie Corelli's; but a book can hardly be entirely bad if it fires a clever boy's imagination. There are two heroes, a primary hero called Flowers, who is almost wholly good, and a secondary hero, a much weaker vessel, called Brown. Flowers and Brown find many dangers in university life, but the worst is a gambling saloon in Chesterton† run by the Misses Bellenden, two fascinating but extremely wicked young ladies. Flowers survives all these troubles, is Second Wrangler and Senior Classic, and succeeds automatically to a Fellowship (as I suppose he would have done then). Brown succumbs, ruins his parents, takes to drink, is saved from delirium tremens during a thunderstorm only by the prayers of

* 'Alan St Aubyn' was Mrs Frances Marshall, wife of Matthew Marshall.

† Actually, Chesterton lacks picturesque features.

the Junior Dean, has much difficulty in obtaining even an Ordinary Degree, and ultimately becomes a missionary. The friendship is not shattered by these unhappy events, and Flowers's thoughts stray to Brown, with affectionate pity, as he drinks port and eats walnuts for the first time in Senior Combination Room.

Now Flowers was a decent enough fellow (so far as 'Alan St Aubyn' could draw one), but even my unsophisticated mind refused to accept him as clever. If he could do these things, why not I? In particular, the final scene in Combination Room fascinated me completely, and from that time, until I obtained one, mathematics meant to me primarily a Fellowship of Trinity.

I found at once, when I came to Cambridge, that a Fellowship implied original work', but it was a long time before I formed any definite idea of research. I had of course found at school, as every future mathematician does, that I could often do things much better than my teachers; and even at Cambridge I found, though naturally much less frequently, that

I could sometimes do things better than the College lecturers. But I was really quite ignorant, even when I took the Tripos, of the subjects on which I have spent the rest of my life; and I still thought of mathematics as essentially a 'competitive' subject. My eyes were first opened by Professor Love, who taught me for a few terms and gave me my first serious conception of analysis. But the great debt which I owe to him—he was, after all, primarily an applied mathematician—was his advice to read Jordan's famous *Cours d'analyse*; and I shall never forget the astonishment with which I read that remarkable work, the first inspiration for so many mathematicians of my generation, and learnt for the first time as I read it what mathematics really meant. From that time onwards I was in my way a real mathematician, with sound mathematical ambitions and a genuine passion for mathematics.

I wrote a great deal during the next ten years, but very little of any importance; there are not more than four or five papers which I can still remember with some satisfaction. The real crises of my career came ten or twelve years later, in 1911, when I began my long collaboration with Littlewood, and in 1913, when I discovered Ramanujan. All my best work since then has been bound up with theirs, and it is obvious that my association with them was the decisive event of my life. I still say to myself when I am depressed, and find myself forced to listen to pompous and tiresome people, 'Well, I have done one thing *you* could never have done, and that is to have collaborated with both Littlewood and Ramanujan on something like equal terms.' It is to them that I owe an unusually late maturity: I was at my best at a little past forty, when I was a professor at Oxford. Since then I have suffered from that steady deterioration which is the common fate of elderly men and particularly of elderly mathematicians. A mathematician may still be competent enough at sixty, but it is useless to expect him to have original ideas.

It is plain now that my life, for what it is worth, is finished, and that nothing I can do can perceptibly increase or diminish its value. It is very difficult to be dispassionate, but I count it a 'success'; I have had more reward and not less than was due to a man of my particular grade of ability. I have held a series of comfortable and 'dignified' positions. I have had very little trouble with the duller routine of universities. I hate 'teaching', and have had to do very little, such teaching as I have done having been almost entirely supervision of research; I love lecturing, and have lectured a great deal to extremely able classes; and I have always had plenty of leisure for the researches which have been the one great permanent happiness of my life. I have found it easy to work with others, and have collaborated on a large scale with two exceptional mathematicians; and this has enabled me to add to mathematics a good deal more than I could reasonably have expected. I have had my disappointments, like any other mathematician, but none of them has been too serious or has made me particularly unhappy. If I had been offered a life neither better nor worse when I was twenty, I would have accepted without hesitation.

It seems absurd to suppose that I could have

'done better'. I have no linguistic or artistic ability, and very little interest in experimental science. I might have been a tolerable philosopher, but not one of a very original kind. I think that I might have made a good lawyer; but journalism is the only profession, outside academic life, in which I should have felt really confident of my chances. There is no doubt that I was right to be a mathematician, if the criterion is to be what is commonly called success.

My choice was right, then, if what I wanted was a reasonably comfortable and happy life. But solicitors and stockbrokers and bookmakers often lead comfortable and happy lives, and it is very difficult to see how the world is the richer for their existence. Is there any sense in which I can claim that my life has been less futile than theirs? It seems to me again that there is only one possible answer: yes, perhaps, but, if so, for one reason only.

I have never done anything 'useful'. No discovery of mine has made, or is likely to make, directly or indirectly, for good or ill, the least difference to the amenity of the world. I have helped to train other mathematicians, but mathematicians of the same kind as myself, and their work has been, so far at any rate as I have helped them to it, as useless as my own. Judged by all practical standards, the value of my mathematical life is nil; and outside mathematics it is trivial anyhow. I have just one chance of escaping a verdict of complete triviality, that I may be judged to have created something worth creating. And that I have created something is undeniable: the question is about its value.

The case for my life, then, or for that of any one else who has been a mathematician in the same sense in which I have been one, is this: that I have added something to knowledge, and helped others to add more; and that these somethings have a value which differs in degree only, and not in kind, from that of the creations of the great mathematicians, or of any of the other artists, great or small, who have left some kind of memorial behind them.

Note

Professor Broad and Dr Snow have both remarked to me that, if I am to strike a fair balance between the good and evil done by science, I must not allow myself to be too much obsessed by its effects on war; and that, even when I am thinking of them, I must remember that it has many very important effects besides those which are purely destructive. Thus (to take the latter point first), I must remember (*a*) that the organization of an entire population for war is only possible through scientific methods; (*b*) that science has greatly increased the power of propaganda, which is used almost exclusively for evil; and (*c*) that it has made 'neutrality' almost impossible or unmeaning, so that there are no longer 'islands of peace' from which sanity and restoration might spread out gradually after war. All this, of course, tends to reinforce the case *against* science. On the other hand, even if we press this case to the utmost, it is hardly possible to maintain seriously that the evil done by science is not altogether outweighed by the good. For example, if ten million lives were lost in every war, the net effect of science would still have been to increase the average length of life. In short, my § 28 is much too 'sentimental'.

I do not dispute the justice of these criticisms, but, for the reasons which I state in my preface, I have found it impossible to meet them in my text, and content myself with this acknowledgement.

Dr Snow has also made an interesting minor point about § 8. Even if we grant that 'Archimedes will be remembered when Aeschylus is forgotten', is not mathematical fame a little too 'anonymous' to be wholly satisfying? We could form a fairly coherent picture of the personality of Aeschylus (still more, of course, of Shakespeare or Tolstoi) from their works alone, while Archimedes and Eudoxus would remain mere names.

Mr J. M. Lomas put this point more picturesquely when we were passing the Nelson column in Trafalgar Square. If I had a statue on a column in London, would I prefer the column to be so high that the statue was invisible, or low enough for the features to be recognizable? I would choose the first alternative, Dr Snow, presumably, the second.

Printed in the United States
By Bookmasters